Lipase- and protease-catalyzed transformations
with unnatural acyl acceptors

Lipase- and protease-catalyzed transformations with unnatural acyl acceptors

PROEFSCHRIFT

ter verkrijging van de graad van doctor,
aan de Technische Universiteit Delft,
op gezag van de Rector Magnificus Prof. ir. K. F. Wakker,
in het openbaar te verdedigen ten overstaan van een commissie,
door het College van Dekanen aangewezen,
op donderdag 29 juni 1995 te 13.30 uur
door

Maria Christina DE ZOETE

doctorandus in de scheikunde
geboren te 's-Gravenhage

Dit proefschrift is goedgekeurd door de promotor:
Prof. dr. R. A. Sheldon

Toegevoegd promotor: Dr. ir. F. van Rantwijk

Samenstelling promotiecommissie:

Rector Magnificus	Technische Universiteit Delft
Prof. dr. R. A. Sheldon	Technische Universiteit Delft
Dr. ir. F. van Rantwijk	Technische Universiteit Delft
Prof. dr. ir. J. J. Heijnen	Technische Universiteit Delft
Prof. dr. A. van der Gen	Rijksuniversiteit Leiden
Prof. dr. B. Zwanenburg	Katholieke Universititeit Nijmegen
Dr. ing. J. Kamphuis	DSM Research

Published and distributed by:

Delft University Press
Stevinweg 1
2628 CN Delft
The Netherlands
Telephone +31 15 783254
Fax +31 15 781661

CIP-DATA KONINKLIJKE BIBLIOTHEEK, DEN HAAG

Zoete, Maria Christina de

Lipase- and protease-catalyzed transformations with unnatural acyl acceptors / Maria Christina de Zoete. - Delft : Delft University Press. - III.
Thesis Delft University of Technology. - With ref. - With summary in Dutch.
ISBN 90-407-1135-6
NUGI 841
Subject headings: biocatalysis / perhydrolysis / ammoniolysis.

aan mijn ouders
voor Mark

CONTENTS

Chapter 1

Lipase-catalyzed transformations with unnatural acyl acceptors

This chapter has been published: M. C. de Zoete, F. van Rantwijk and R. A. Sheldon, *Catalysis Today*, **22**, 563 (1994).

INTRODUCTION

Lipases (triacylglycerol acylhydrolase EC 3.1.1.3) belong to the class of serine hydrolases[1]. They occur widely in nature where their biological function is to catalyze the hydrolysis of triglycerides to the corresponding fatty acids and glycerol (reaction 1). Lipases are active at oil-water interfaces, in contrast to other carboxylesterases which function optimally with water-soluble substrates in homogeneous media.

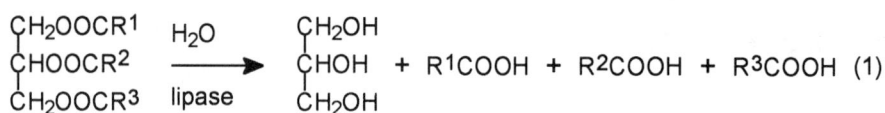

$$\begin{array}{l} CH_2OOCR^1 \\ CHOOCR^2 \\ CH_2OOCR^3 \end{array} \begin{array}{c} H_2O \\ \longrightarrow \\ lipase \end{array} \begin{array}{l} CH_2OH \\ CHOH \\ CH_2OH \end{array} + R^1COOH + R^2COOH + R^3COOH \quad (1)$$

1.1. Why are lipases interesting catalysts?

One reason which makes lipases interesting catalysts for industrial organic synthesis is their availability. In recent years lipases of microbial origin have been widely applied in the dairy and food industries (cheese ripening and the production of cocoa butter substitutes), laundry detergents, tanning and cosmetics[2]. Hence, a number of microbial lipases, e.g. from *Candida*, *Rhizomucor* and *Pseudomonas* species, produced using recombinant DNA techniques, have become readily available for use in organic synthesis (see table 1 for an overview of commercially available lipases). This is reflected in the trend in the literature away from the use of impure isolates of porcine pancreatic lipase (PPL), which are contaminated with other enzymes, to the use of high purity microbial lipases.

Table 1. *Overview of commonly used commercially available lipases*

Lipase	Supplier	Trade name
C. rugosa[*]	Sigma, Amano, Meito Sangyo Co.	Amano R, Amano AY, lipase OF360
C. antarctica	Novo Nordisk A/S	Novozyme 435
Rh. miehei[**]	Novo Nordisk	Lipozyme
Humicola	Novo Nordisk A/S	
PPL	Sigma	
Pancreatin	Sigma	
Pseudomonas	Amano	Amano PS, Amano AK
Rhizopus arrhizus	Boehringer Mannheim	
Pseudomonas lipoprotein lipase	Boehringer Mannheim	

[*] The name of *Candida cylindracea* has been changed to *Candida rugosa*.
[**] The name of *Mucor miehei* has been changed to *Rhizomucor miehei*.

Another beneficial feature of lipases is their excellent stability. They are robust enzymes that can tolerate relatively hostile conditions, e.g. in organic solvents at elevated temperatures. *Candida antarctica* lipase, for example, functions admirably at 80°C in hydrophobic solvents[3]. A third characteristic feature of lipases is their promiscuous acceptance of a wide variety of unnatural substrates. They have a rather shallow, hydrophobic active site, which is reflected in a preference for linear carboxylic acid moieties, analogous to their natural substrates. Much more variation is tolerated, however, in the alcohol moiety of the substrate ester.

Last, but not least, lipases are able to catalyze the enantioselective hydrolysis of chiral carboxylic esters. Taken together with their ability to accept a wide variety of unnatural acyl acceptors and donors this renders them broadly applicable in organic synthesis. Consequently, lipases have been widely applied in recent years in the commercial synthesis of optically pure pharmaceuticals, agrochemicals, flavors and food additives[4,5a].

1.2. Examples of industrial applications

Some examples of commercial applications of lipase-catalyzed transformations are shown in reactions 2-5. The resolution of α-bromopropionic acid (reaction 2), to give an intermediate for the synthesis of optically pure (R)-α-phenoxypropionic acid herbicides, has been commercialized by Chemie Linz in Austria[5]. The resolutions of glycidyl butyrate (reaction 3) and p-methoxyphenylglycidate ester (reaction 4) have been commercialized by DSM-Andeno[6]. The products are used as intermediates in the preparation of optically active beta-blockers[6a] and the anti-hypertensive drug, Diltiazem[6b], respectively. Another application involves the use of lipases for the selective acylation of ethyl glucoside at the 6-position (reaction 5). The product is used in laundry detergent applications[7].

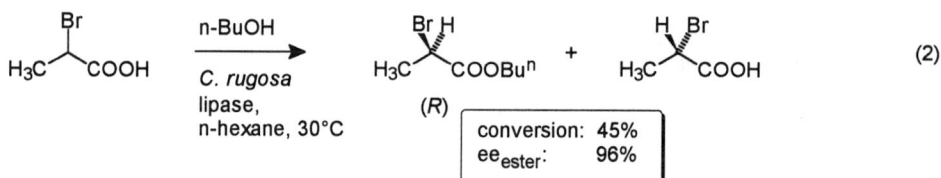

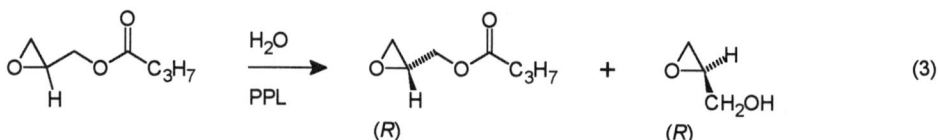

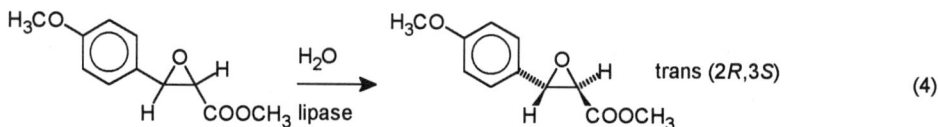

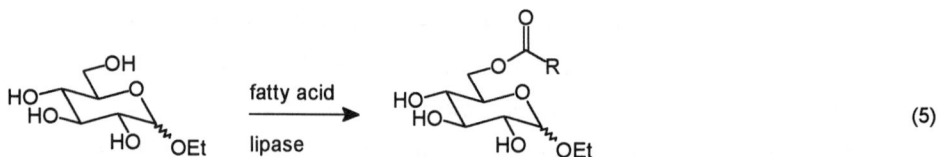

1.3. Mechanism of lipase-catalyzed reactions

As noted earlier lipases belong to the group of serine hydrolases. A common feature of these enzymes is the involvement of a serine-histidine-aspartate triad in the catalytic mechanism, as illustrated in Figure 1. The acyl-enzyme intermediate is formed by reaction of the substrate with the OH group of a serine residue, which is assisted by the His and Asp groups. Subsequent reaction of the acyl-enzyme intermediate with a nucleophile affords the product. In the case of a hydrolysis the nucleophile is water whilst in an esterification it is an alcohol. However, any nucleophile can, in principle, react with the acyl-enzyme intermediate and, as we shall see later, this significantly broadens the synthetic scope of lipases.

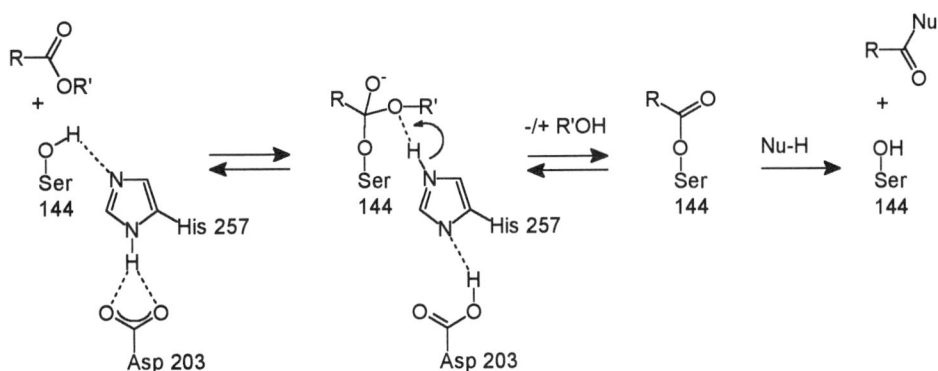

Numbering is for Rhizomucor miehei lipase

------- denotes hydrogen bond

Figure 1. *Reaction mechanism of lipase-catalyzed reactions.*

5

Obviously, in aqueous media ester hydrolysis (reaction 6) is observed. In non-aqueous media, on the other hand, it is possible to carry out esterifications (reaction 7), transesterifications (reaction 8) and interesterifications (reaction 9).

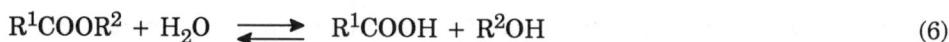

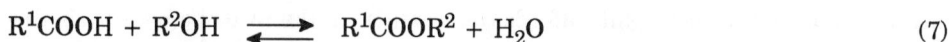

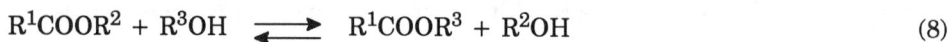

$$R^1COOR^2 + H_2O \rightleftharpoons R^1COOH + R^2OH \qquad (6)$$

$$R^1COOH + R^2OH \rightleftharpoons R^1COOR^2 + H_2O \qquad (7)$$

$$R^1COOR^2 + R^3OH \rightleftharpoons R^1COOR^3 + R^2OH \qquad (8)$$

$$R^1COOR^2 + R^3COOR^4 \rightleftharpoons R^1COOR^4 + R^3COOR^2 \qquad (9)$$

In these transformations water acts as a competing nucleophile. The amount of competition is dependent on the thermodynamic water *activity* of the system[8] and not on the water *concentration* in the organic phase or the total water content. To avoid this unwanted hydrolysis, it is necessary to work at very low water activity, which can be effected by exhaustive drying. However, a number of lipases (such as *Rh. miehei* lipase) are less active at low water activity[9] and for these enzymes it might be advantageous to use salt hydrates[10] to fix the water activity at the optimum value. On the contrary, other lipases, such as *C. antarctica* lipase, are even more active and selective at very low water activity[9]. With these enzymes hydrolysis can be excluded completely.

In the context of practical applications it is worth noting that immobilization of the lipase, on acrylic polymers, ion-exchange resins or polypropylene (Accurel EP100), not only facilitates their recovery but also enhances their stability in organic media. In addition, enzymes can also be immobilized in membranes for use in membrane reactors[11]. Using this technique enzyme stabilisation is obtained because product inhibition and enzyme deactivation are minimized as a result of feed dosing and fast extraction of the products from the reaction mixture.

1.4 Kinetic considerations[12]

The kinetics of irreversible enzymatic reactions, such as hydrolysis of esters in water, are effectively described by Michaelis-Menten kinetics in which two enantiomeric substrates are converted to the corresponding enantiomeric products with different reaction rates. The velocity of the transformation of each enantiomer is dependent on the conversion, because the composition of the enantiomeric mixture does not remain constant during the reaction. As a result thereof, the enantiomeric excess of the substrate and product is dependent on the conversion (see Figure 2).

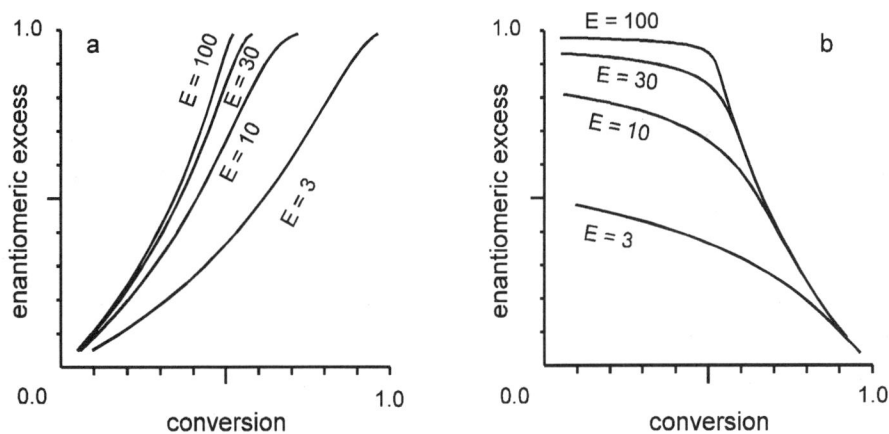

Figure 2. *Dependence of enantiomeric excess of a) the substrate and b) the product on the conversion in the case of an irreversible reaction.*

The enantiomeric ratio (E), which remains constant throughout the reaction was introduced to describe the selectivity of the reaction. E is mathematically linked to the conversion (c) and the optical purities of substrate (ee_s) and product (ee_p) and can be expressed as follows:

For the substrate:

$$E = \frac{\ln[(1-c)(1-ee_s)]}{\ln[(1-c)(1+ee_s)]}$$

For the product:

$$E = \frac{\ln[1-c(1+ee_p)]}{\ln[1-c(1-ee_p)]}$$

When the reaction is reversible the situation becomes more complex. In this case the equilibrium constant ($K = (1 - c_e) / c_e$; c_e = conversion at equilibrium) is an important parameter. When the reverse reaction starts to take place to a significant extent (which is dependent on the position of the equilibrium) the enantiomeric purity of both the substrate and product is decreased (see Figure 3). To avoid this it is necessary to shift the equilibrium by using about 20 molar equivalents of nucleophile vs. substrate.

The dependence of the enantiomeric ratio (E) on the conversion (c) and the optical purities of substrate (ee_s) and product (ee_p) in this case is as follows:

For the substrate:

$$E = \frac{\ln[1-(1+K)(c+ee_s(1-c))]}{\ln[1-(1+K)(c-ee_s(1-c))]}$$

For the product:

$$E = \frac{\ln[1-(1+K)c(1+ee_p)]}{\ln[1-(1+K)c(1-ee_p)]}$$

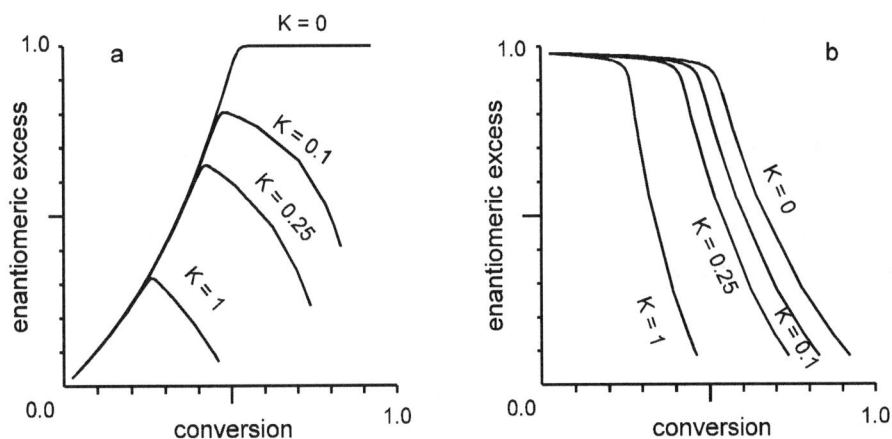

Figure 3. *Dependence of the enantiomeric excess on the conversion of a) the substrate and b) the product in the case of a reversible reaction.*

Recently, a new simple method to determine the enantiomeric ratio (E) directly from the enantiomeric excess of the substrate (ee_s) and product (ee_p) was described[13]. The following equation was obtained:

$$E = \frac{\ln\left[\dfrac{(1-ee_s)}{(1+\dfrac{ee_s}{ee_p})}\right]}{\ln\left[\dfrac{1+ee_s)}{(1+\dfrac{ee_s}{ee_p})}\right]}$$

9

2. HYDROLYSIS

2.1. Carboxyl esters

Lipases have been shown to catalyze the enantioselective hydrolysis of a wide variety of unnatural ester substrates. Two examples are illustrated in reactions 10[14] and 11[15].

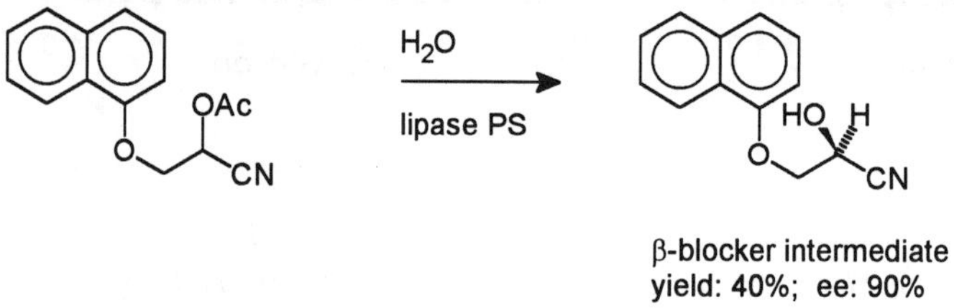

(10)

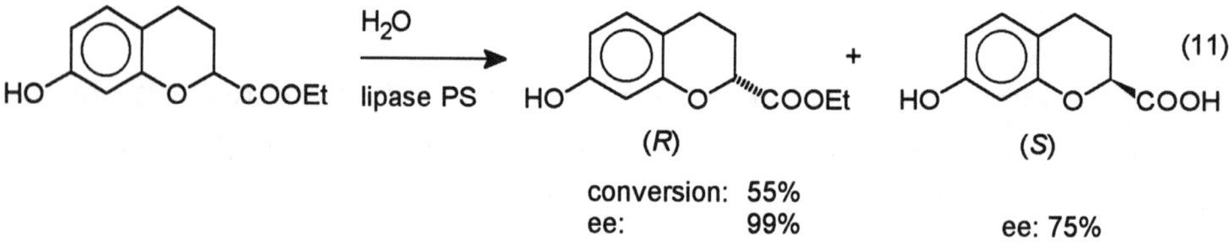

β-blocker intermediate
yield: 40%; ee: 90%

(11)

conversion: 55%
ee: 99%

ee: 75%

The lipase-catalyzed hydrolysis of amino acid esters has been used for the selective removal of protecting groups in peptides and glycopeptides, e.g.[16],

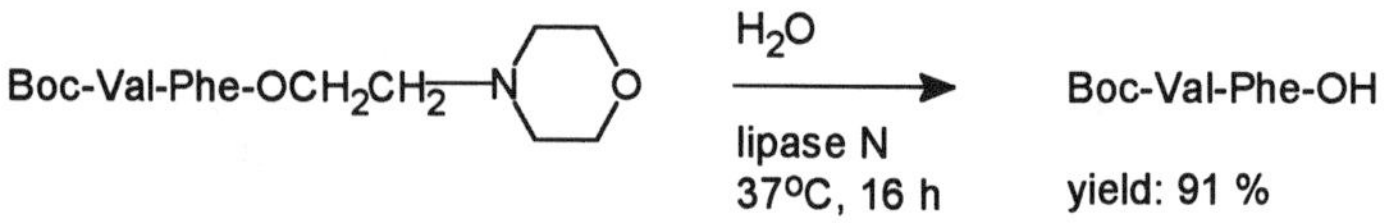

Boc-Val-Phe-OCH$_2$CH$_2$-N(morpholine) $\xrightarrow[\substack{\text{lipase N} \\ 37^{\circ}C, 16\,h}]{H_2O}$ Boc-Val-Phe-OH   (12)

yield: 91 %

2.2. Thiol esters

Kinetic experiments revealed that lipases accept thiol esters as substrates (reaction 13 and 14) although rates are generally lower compared to esters of alcohols[17].

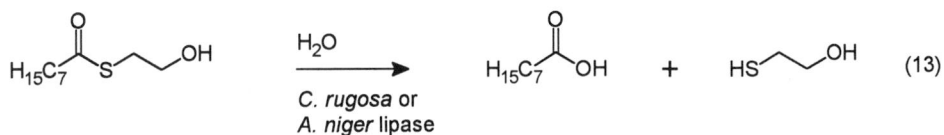

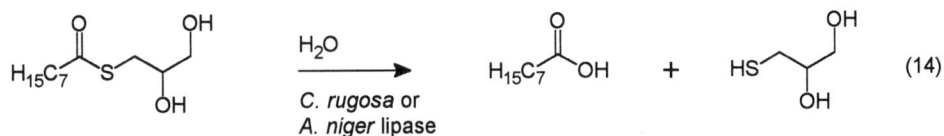

$$ (13) $$

$$ (14) $$

3. ESTERIFICATION AND TRANSESTERIFICATION

3.1. Chiral alcohols

In organic media alcohols can be acylated with a carboxylic acid (esterification) or an ester (transesterification, e.g.[18]).

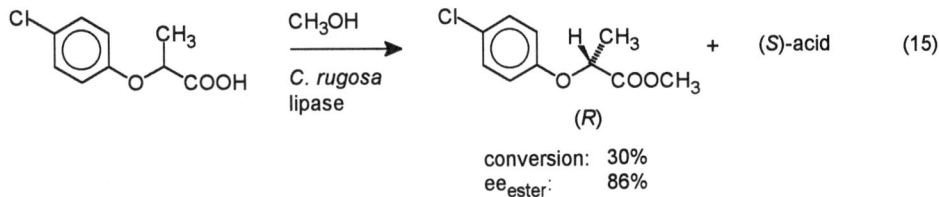

conversion: 30%
ee_{ester}:   86%

$$ (15) $$

11

The structure of the acyl group can have a significant effect on the enantioselectivity (defined by the enantiomeric ratio E) observed in acylations of chiral alcohols, e.g.[19].

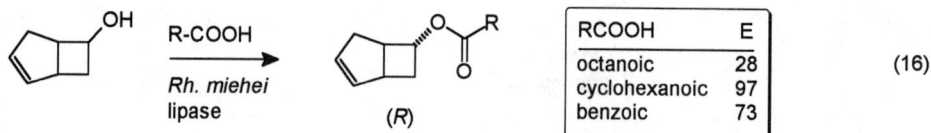

RCOOH	E
octanoic	28
cyclohexanoic	97
benzoic	73

(16)

Since such esterifications are reversible, product yields can generally be improved by removal of the water coproduct, e.g. with molecular sieves. Alternatively, if a transesterification is performed the equilibrium can be shifted to the right if the alcohol coproduct can be continuously removed by distillation, i.e. if there is sufficient difference in boiling points of the alcohol substrate and alcohol product, e.g. as in the example shown (reaction 17) [20].

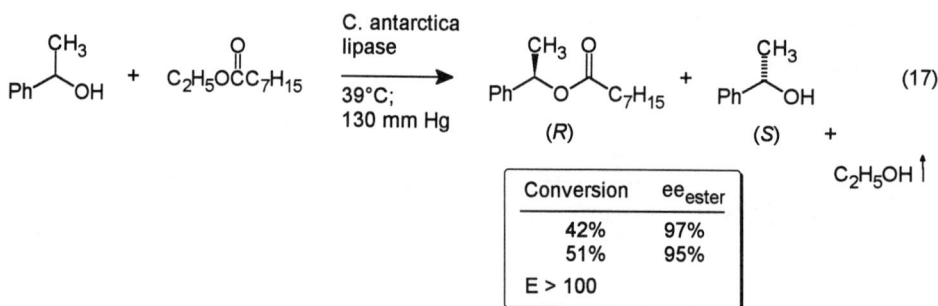

(17)

Conversion	ee_{ester}
42%	97%
51%	95%
E > 100	

The inherent reversibility of transesterification can lead to a loss of enantioselectivity in acylations of chiral alcohols. Thus, if acylation of a racemic alcohol affords predominantly the (*R*)-ester (reaction 18) then the reverse reaction of the achiral product (R^3OH) with the (*R*)-ester product will afford the (*R*)-alcohol. In other words, the forward reaction affords the (*S*)-alcohol and the reverse reaction yields the (*R*)-alcohol.

$$R^1OH + R^2COOR^3 \rightleftharpoons R^1OH + R^2COOR^1 + R^3OH \qquad (18)$$

(*R*,*S*) (*S*) (*R*)

One way of circumventing this problem is to use a large excess of acyl donor, e.g. a commonly employed method is to use ethyl acetate as both the acyl donor and the solvent for lipase-catalyzed acylations of chiral alcohols[21]. Alternatively, acyl donors with activated leaving groups, e.g. trichloroethyl, trifluoroethyl or oxime esters[22] can be used:

CH3
X
OH
(R,S)
+
Cl3C
O
O
C3H7
PPL,
hexane or
Et2O
CH3
X
OH
(S)
+
CH3
O
X
O
C3H7
(R)
eealc. 90-100%
eeester 90-100%
X = -n-C6H13
-n-C10H21
-n-C14H29
-CH2-C(OH)(CH3)2
-C6H5

(19)

O
O
H3C
O
N
CH3
CH3
+ HO
O
H3C
O
conv. 90% in 4 h
+
O
HO
N
CH3
CH3

(20)

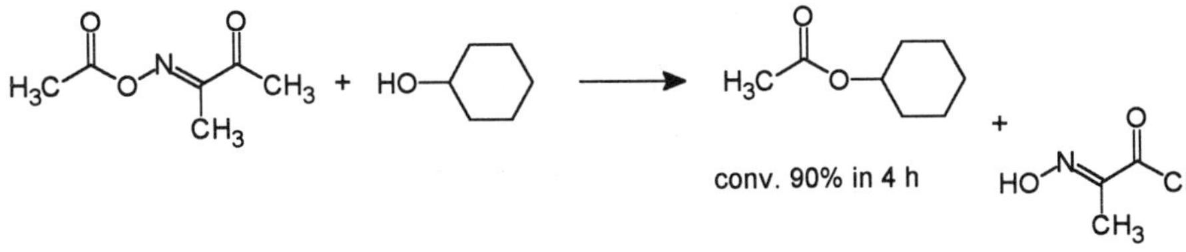

Another method of avoiding this problem is to use a vinylic ester as acyl donor. In this case the alcohol product spontaneously tautomerizes to the corresponding carbonyl compound, thus rendering the reaction irreversible, e.g.[23].

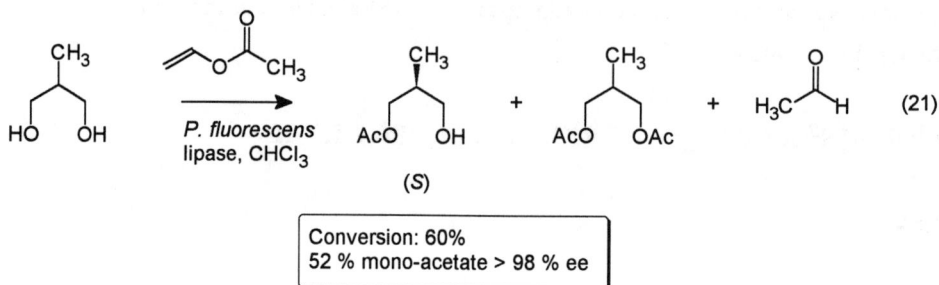

However it should be noted that the liberated acetaldehyde can cause deactivation of the lipase by reaction with lysine residues[24].

Another variation on this theme is the use of acid anhydrides[25] as acyl donors (reaction 22). The reaction is highly selective but the liberated acid can lower the pH of the medium which can lead to deactivation. Moreover, the blank reaction may be significant with acid anhydrides.

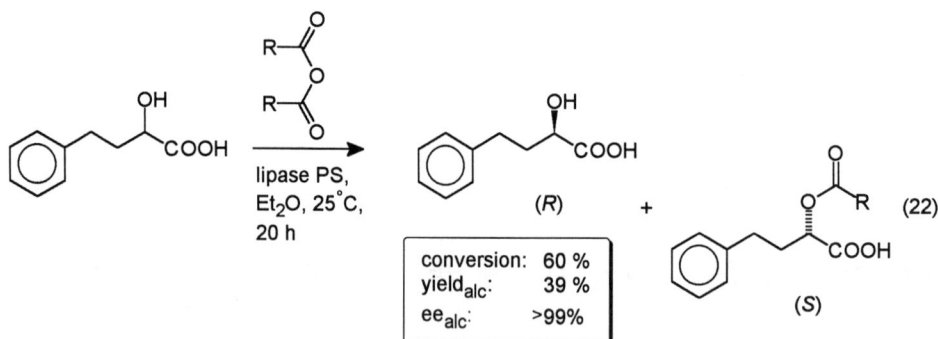

14

3.2. Thiol esters

Interestingly, racemic ß-acetylmercaptoisobutyric acid undergoes chemoselective thiotransesterification with n-propanol (reaction 23) in the presence of PPL or lipase P[26].

Enzyme	(S)-ester		
	conversion (%)	yield(%)	ee(%)
Lipase P	60	31	93
PPL	51	42	95

3.3. Unusual alcohols

Alcohols which contain silicon are also accepted as acyl donors in lipase catalyzed transesterifications (reaction 24). The silicon atom closely resembles a carbon atom and enhanced both the reaction rate and enantioselectivity of the transesterification[27].

Another class of unusual alcohols which can be used in enantioselective transesterifications are alcohols containing metal elements such as ferrocenes, stannanes and chromium compounds[28]. Examples of these alcohols are shown in Figure 4.

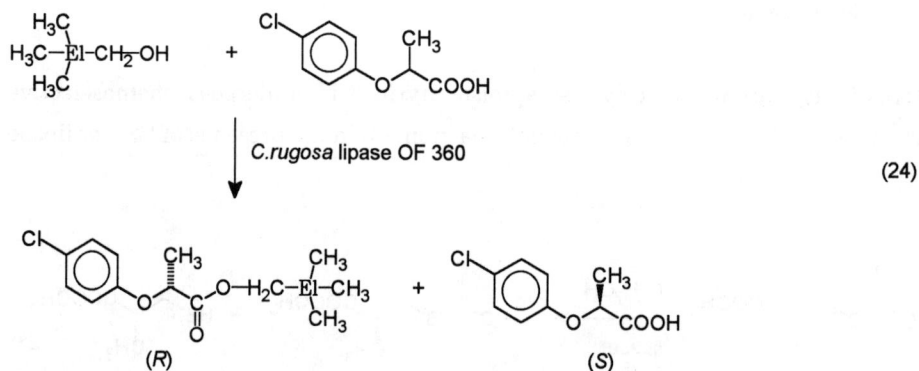

C.rugosa lipase OF 360

(24)

(R)   +   (S)

El	Reaction time (h)	Conversion (%)	ee$_{acid}$ (%)
Si	19	51.7	95.8
C	196	52.8	91.1

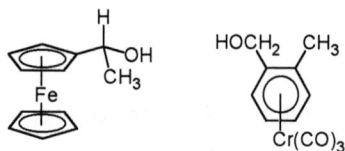

Figure 4. *Examples of alcohols containing metal elements.*

16

4. PROCHIRAL AND MESO DIOLS AS ACYL ACCEPTORS

In the lipase-mediated kinetic resolutions of chiral alcohols described in sections 2 and 3 the maximum theoretical yield of a pure enantiomer is, by definition, 50%. When prochiral or meso diols are used as the acyl acceptors, in contrast, one can obtain complete conversion of the substrate to one enantiomer. Typical examples of acylations of a prochiral diol[29] and a meso diol[30] are illustrated in reactions 25 and 26, respectively.

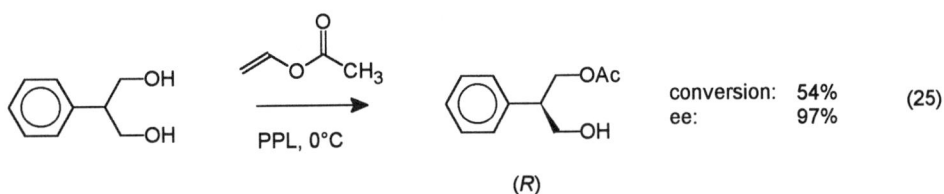

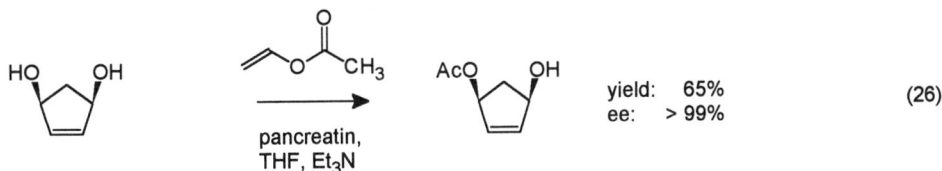

The enantioselectivity of such reactions of prochiral substrates is markedly influenced by both the particular lipase used and the solvent, as illustrated in reactions 27 and 28, respectively[31].

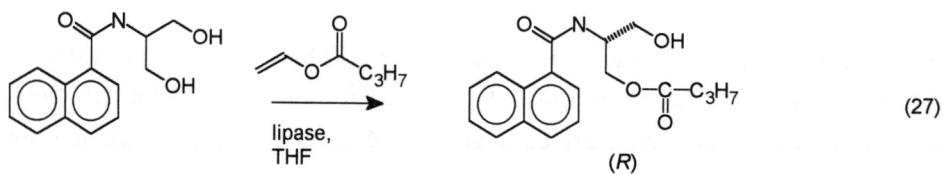

(27)

Lipase source	E
Pseudomonas sp.	> 27
porcine pancreas	> 27
Chromobacterium viscosum	8
Rhizopus arrhizus	4

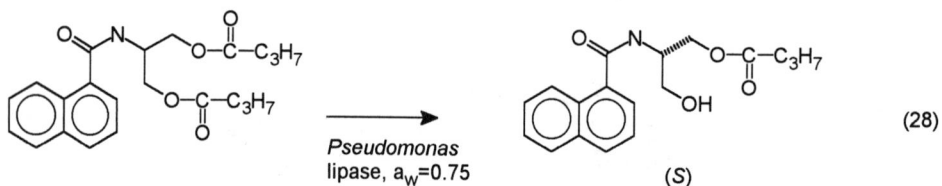

(28)

Solvent	E
acetonitrile	> 30
nitrobenzene	> 30
acetone	18
chloroform	10
dioxane	9
t-butyl alcohol	5
tetrachlorocarbon	3

5. SUGARS AS ACYL ACCEPTORS

Carbohydrates comprise a special class of alcohols that can act as acyl acceptors in lipase-catalyzed transesterifications. The resulting acylated carbohydrates have commercial potential as biodegradable surfactants, bioemulsifiers and liquid crystals[32].

5.1. Monosaccharides

The first enzymatic acylation of sugars was reported by Klibanov[32], who showed that glucose is selectively acylated at the 6-position using trichloroethyl esters with PPL in pyridine (reaction 29).

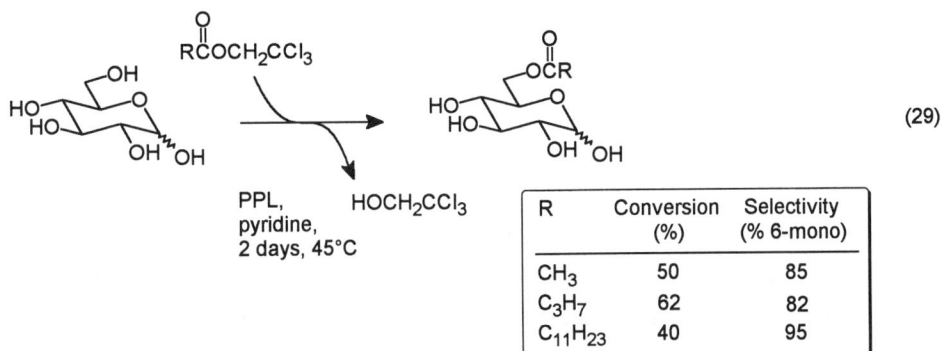

R	Conversion (%)	Selectivity (% 6-mono)
CH$_3$	50	85
C$_3$H$_7$	62	82
C$_{11}$H$_{23}$	40	95

(29)

The acylation of D-fructose with fatty acids (reaction 30a) has been reported in the more environmentally acceptable 2-methyl-2-butanol as solvent[34]. Alternatively, the acylation could be performed in n-hexane (reaction 30b) when phenylboronic acid was added to enhance the solubility of D-fructose by complex formation[35]. In the latter case a mixture of isomers was obtained.

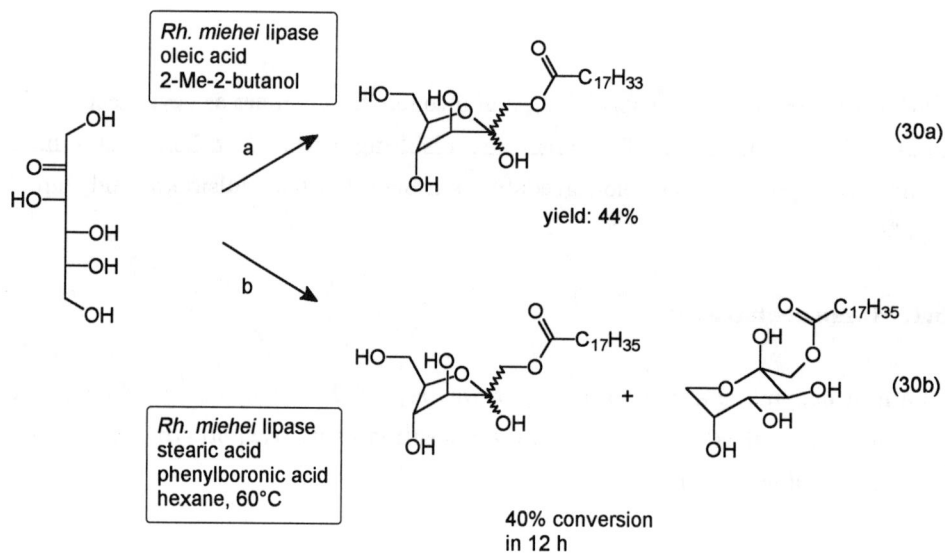

(30a)

yield: 44%

(30b)

40% conversion
in 12 h

More recently, we have shown[36] that glucose is acylated with ethyl esters, using *Candida antarctica* lipase, in *t*-butyl alcohol or in the absence of a solvent (reaction 31). When no solvent is used the reaction rate is low and diesters are preferentially formed. In *t*-butyl alcohol, on the other hand, the rate is higher and preferential formation of the mono ester at the primary (6)-position is observed.

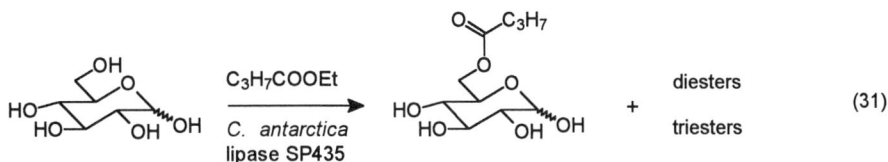

$$(31)$$

Solvent	Products in %		
	mono-6	di	tri
Ethyl butanoate	2	13	3
t-BuOH	48	5	0

Conditions: 40 mg glucose, 40 mg *C. antarctica* lipase SP 435, 4 ml ethyl butanoate or 4 ml 0.94 M ethyl butanoate in *t*-butyl alcohol, 0.4 g zeolite CaA, 40 °C, 48 h.

5.2. Disaccharides

Klibanov and coworkers[37] reported the acylation of sucrose at the 1'-position with the activated trichloroethyl butyrate using subtilisin in dimethylformamide as solvent (reaction 32a). More recently we have shown[36] that *Candida antarctica* lipase catalyzed the esterification of sucrose in refluxing *t*-butyl alcohol as solvent (reaction 32b). An approximately 1:1 mixture of 6- and 6'-acylsucrose was obtained; at prolonged reaction time the 6,6'-diacylsucrose was also formed.

Similarly, other disaccharides such as palatinose, leucrose, maltose and trehalose are acylated using *Candida antarctica* lipase in refluxing *t*-butyl alcohol[38].

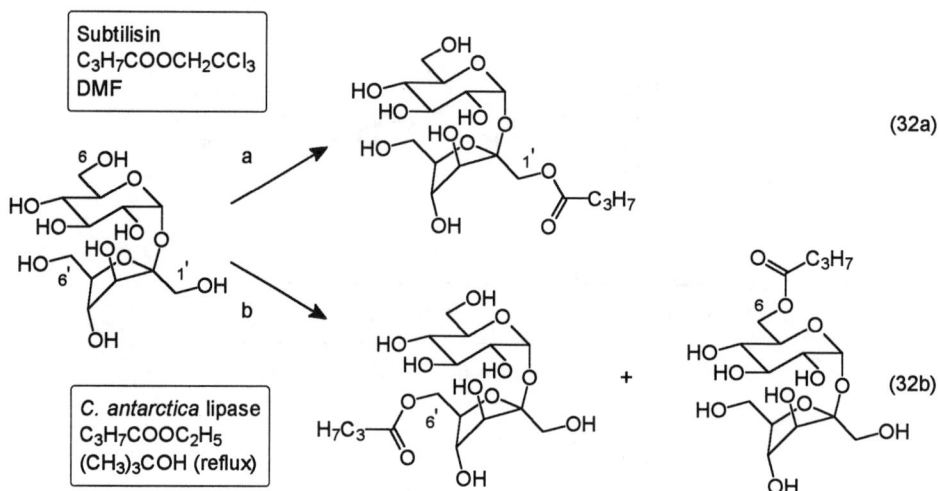

(32a)

(32b)

5.3. Alkyl glucosides and fructosides

Acylated alkyl glucosides have attracted much attention recently because of their interesting surfactant properties[39]. Here again *Candida antarctica* lipase has proven to be a superior catalyst (reaction 33).

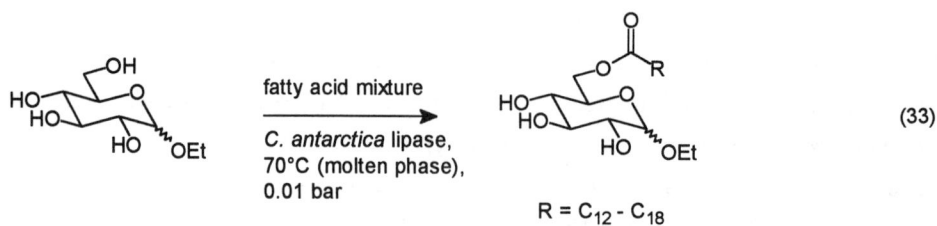

(33)

$R = C_{12} - C_{18}$

Alkyl α-glucosides were selectively acylated with ethyl esters which functioned as both acyl donor and solvent[40]. The reaction rate and selectivity were enhanced by removing the ethanol product using a CaA zeolite. Similarly, α-octyl glucoside was selectivily acylated at the 6-position using ethyl acrylate as the acyl donor[40] (reaction 34). Polymerization of the product affords polymers with interesting properties[41].

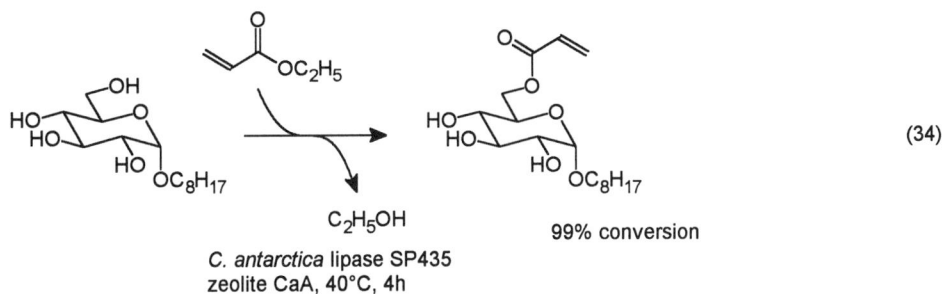

(34)

C. antarctica lipase SP435
zeolite CaA, 40°C, 4h

99% conversion

ß-Dodecyl fructopyranoside is similarly acylated at the primary hydroxyl function using *Candida antarctica* lipase[42].

6. H₂O₂ AND RO₂H AS ACYL ACCEPTORS

Hydrogen peroxide can also act as an acyl acceptor with lipases[43,44]. This leads to the equilibrium formation of peroxycarboxylic acids from the corresponding carboxylic acids. In the presence of a substrate, e.g. an olefin or a dialkylsulfide, which is oxidized by the peroxycarboxylic acid the carboxylic acid is regenerated. Hence, olefin epoxidations and sulfoxidations can be performed, using hydrogen peroxide and a catalytic amount of a carboxylic acid, via *in situ* formation of the peroxycarboxylic acid (reaction 35). As would be expected, long chain acids e.g. octanoic acid, are preferred. Not all lipases tolerate the relatively high concentrations of hydrogen peroxide involved (1.1 mol/l) but a number of lipases, *Rh. miehei*, Amano PS, *Humicola*, and especially *Candida antarctica*, show catalytic activity.

$$H_2O_2 \longrightarrow H_2O$$

$$R\overset{O}{\overset{\|}{C}}OH \qquad R\overset{O}{\overset{\|}{C}}OOH$$

$$SO \longleftarrow S$$

(35)

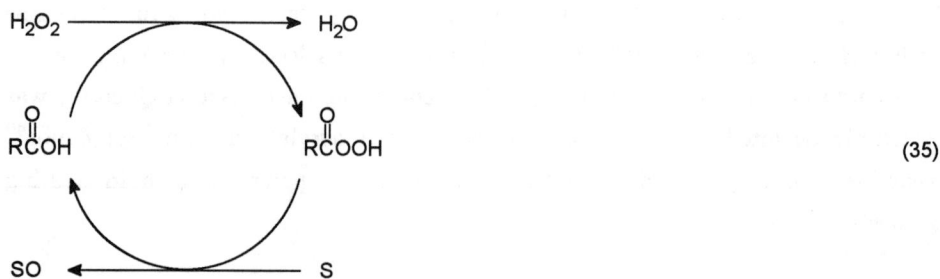

This technique has been used[45] for the selective oxidation of penicillin G to pen-G-sulfoxide (reaction 36).

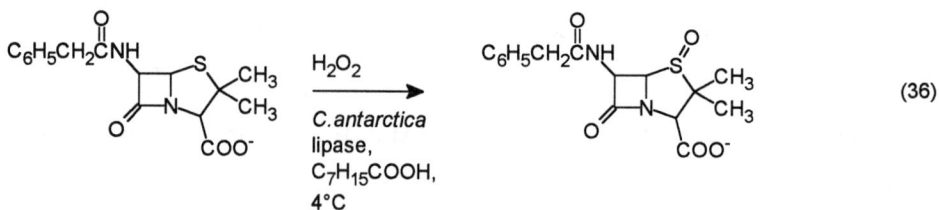

(36)

Alkyl hydroperoxides also act as acyl acceptors and this has been utilized for the kinetic resolution of chiral hydroperoxides[46] using, inter alia, lipoprotein lipase (reaction 37).

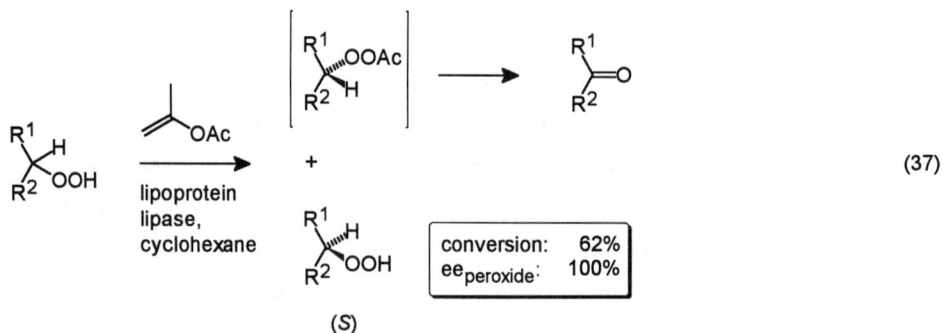

| conversion: | 62% |
| $ee_{peroxide}$: | 100% |

(37)

7. AMINES AS ACYL ACCEPTORS

Amines can also function as acyl acceptors[47-52] as illustrated in reactions 38[53] and 39[54] and this can be utilized for the resolution of chiral esters.

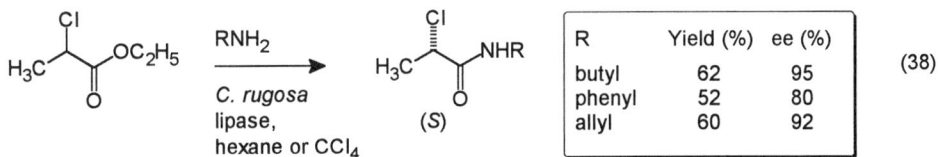

R	Yield (%)	ee (%)
butyl	62	95
phenyl	52	80
allyl	60	92

(38)

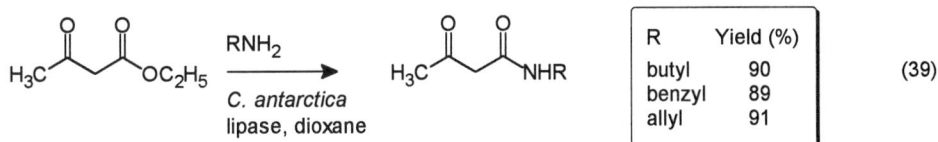

R	Yield (%)
butyl	90
benzyl	89
allyl	91

(39)

The reaction can also be used for the resolution of amines[47] and chiral amino alcohols[55,56] as illustrated in reactions 40 and 41.

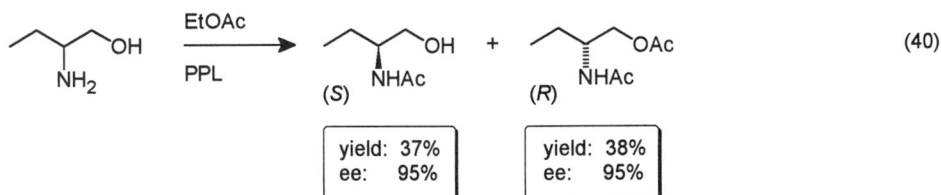

(40)

yield: 37%
ee: 95%

yield: 38%
ee: 95%

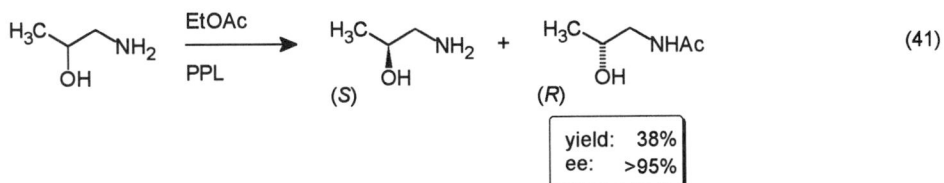

(41)

yield: 38%
ee: >95%

Other authors have reported[57] the lipase-catalyzed transesterification of the acylated derivatives (reaction 42) albeit with modest enantioselectivity.

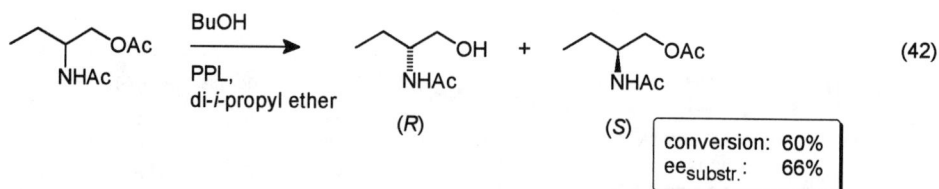

(42)

conversion: 60%
$ee_{substr.}$:   66%

When a racemic ester is used in conjunction with a racemic amine a mixture of diastereomers is obtained. For example, in reaction 43 *Candida antarctica* lipase displayed high enantioselectivity towards the ester component (ee 95%) but rather low enantioselectivity with respect to the amine[58].

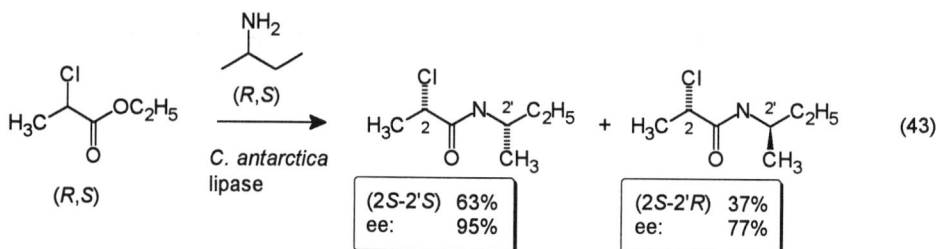

(43)

(2S-2'S) 63%
ee: 95%

(2S-2'R) 37%
ee: 77%

When amino esters are used as substrates intramolecular aminolysis is observed[59], leading to the formation of lactams (reaction 44).

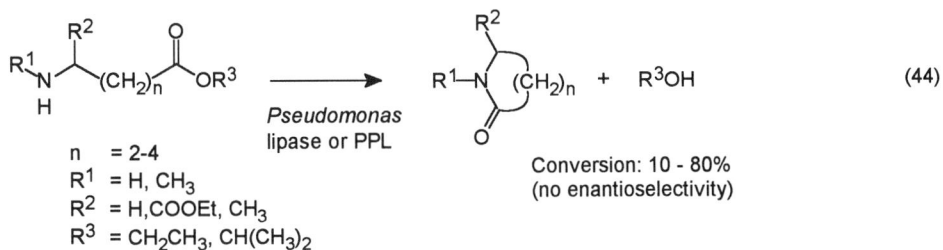

$$n = 2\text{-}4$$
$$R^1 = H, CH_3$$
$$R^2 = H, COOEt, CH_3$$
$$R^3 = CH_2CH_3, CH(CH_3)_2$$

Pseudomonas lipase or PPL

(44)

Conversion: 10 - 80%
(no enantioselectivity)

The reaction of diesters with diamines[59] affords bis-lactams (45).

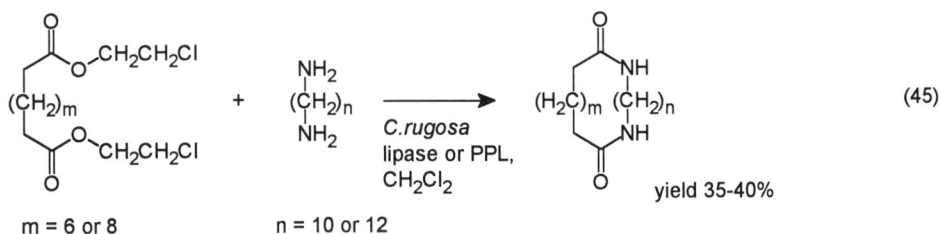

m = 6 or 8   n = 10 or 12   yield 35-40%

(45)

Carboxylic amides generally react sluggishly as acyl donors if at all. The lipase-mediated transamidation of an activated amide[53] is an exception (reaction 46). However, the chemical and optical yields were lower than in the corresponding ester aminolysis.

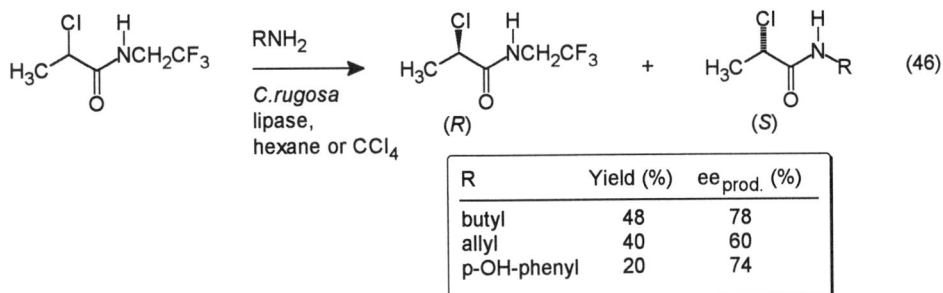

(46)

R	Yield (%)	$ee_{prod.}$ (%)
butyl	48	78
allyl	40	60
p-OH-phenyl	20	74

8. AMMONIA AS ACYL ACCEPTOR

We recently found that ammonia is also an excellent acyl acceptor for certain lipases[60-62]. t-Butyl alcohol is a convenient solvent for enzymatic ammoniolysis as it does not act as an acyl acceptor due to steric hindrance. In ammoniolysis of ethyl octanoate (reaction 47) at 40°C in t-butyl alcohol lipase SP435 from *Candida antarctica*, lipase SP398 from *Humicola sp.*, lipoprotein lipase from *Pseudomonas sp.* and *Rhizopus arrhizus* were able to tolerate high concentrations (2.5 M) of ammonia. Lipases from *Rhizomucor miehei*, *Amano PS*, and PPL displayed ammoniolytic activity only at lower (0.5 M) ammonia concentrations (Table 2).

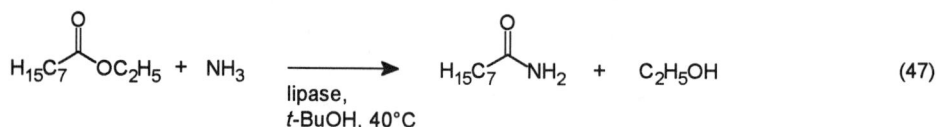

$$H_{15}C_7\text{-}OC_2H_5 + NH_3 \xrightarrow[\text{t-BuOH, 40°C}]{\text{lipase,}} H_{15}C_7\text{-}NH_2 + C_2H_5OH \qquad (47)$$

Table 2. *Ammoniolysis of ethyl octanoate catalyzed by lipases.*

Lipase	% Octanamide in 24 h	
	5 eq. NH$_3$	1 eq. NH$_3$
C. antarctica	95	43
Humicola sp.	85	38
Ps.lipoprotein lipase	15	34
Amano PS	0	7
Rh. miehei	0	20
PPL	0	12

The corresponding carboxylic acids do not undergo ammoniolysis since they form the ammonium salt which cannot act as a substrate for the lipase. However, since the lipases also mediate the esterification of the carboxylic acid the latter can be converted to the amide in a two-step process using the same lipase to catalyze the formation of the ester substrate and its subsequent ammoniolysis[62]. Hence, enzymatic ammoniolysis is a synthetically very useful method for the formation of amides under mild conditions. It could be useful, for example, in the formation of thermally labile amides where the forcing conditions required for the conventional amide synthesis present problems. Moreover, many amides are sparingly soluble in organic solvents and precipitate during the reaction, thus facilitating their isolation. Enzymatic ammoniolysis is also an excellent method for the kinetic resolution of chiral carboxylic acids or chiral alcohols. For example, SP435-mediated ammoniolysis of the 2-chloroethyl ester of ibuprofen (reaction 48) afforded the amide of *(S)*-ibuprofen in excellent enantioselectivity[60].

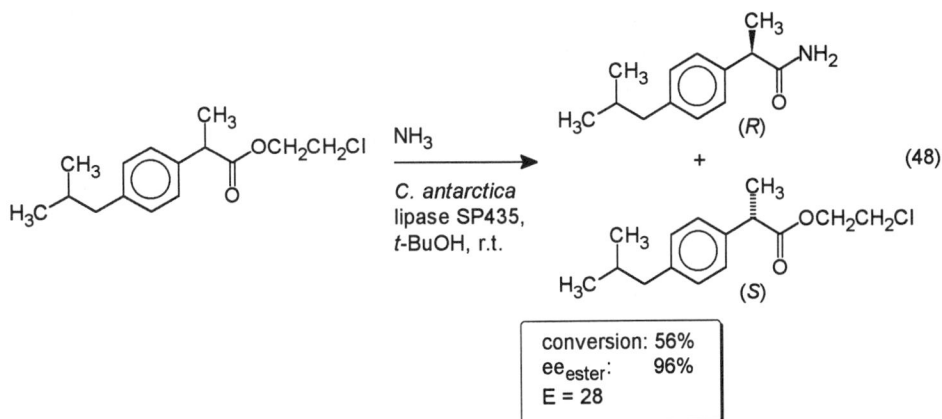

conversion: 56%
ee_{ester}: 96%
E = 28

(48)

Interestingly, the enantioselectivity of the ammoniolysis (E=28) was ca. ten times higher than the corresponding ester hydrolysis (E=4) with the same lipase.

Similarly, enzymatic ester ammoniolysis with SP435 was used for the kinetic resolution of α-methylbenzyl alcohol (reaction 49). In this case high enantioselectivity was observed in both the ammoniolysis and transesterification[62]. The enantioselectivity is presumably determined by discrimination of the ester enantiomers by the enzyme and not by the nature of the acyl acceptor, unless interaction with ammonia results in conformational changes at the active site.

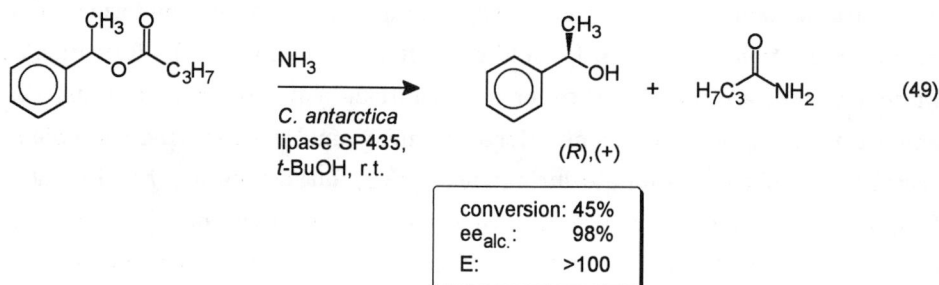

$$(49)$$

conversion: 45%
$ee_{alc.}$: 98%
E: >100

We have also used lipase-mediated ammoniolysis for the conversion of amino acid esters to the corresponding amides[63]. The ethyl ester of racemic phenylglycine, for example, was converted to (*R*)-phenylglycine amide (reaction 50) with a high enantioselectivity (E=38). Moreover, the remaining (*S*)-ester racemises under the reaction conditions used, by which means, in principle, both enantiomers of the ester can be converted into the (*R*)-amide. (*R*)-Phenylglycine and its derivatives are key intermediates in the industrial synthesis of semi-synthetic penicillins and cefalosporins.

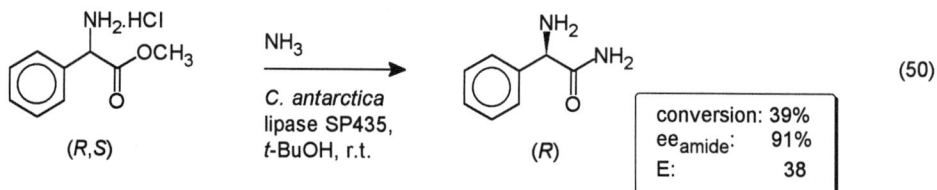

$$(50)$$

conversion: 39%
ee_{amide}: 91%
E: 38

The scope of enzymatic ammoniolysis appears to be broad and we are currently investigating the use of prochiral and meso diesters as substrates, in order to achieve complete conversion to a single enantiomer.

9. HYDRAZINES AS ACYL ACCEPTORS

Alkyl substituted hydrazines have been successfully used as acyl acceptors in reactions mediated by *Pseudomonas cepacia* lipase[64-67]. The best results were obtained with hydrazines containing electron-withdrawing substituents[67]. Chiral esters can be kinetically resolved as shown in reaction 51. The resolution of chiral hydrazines using this technique was not reported.

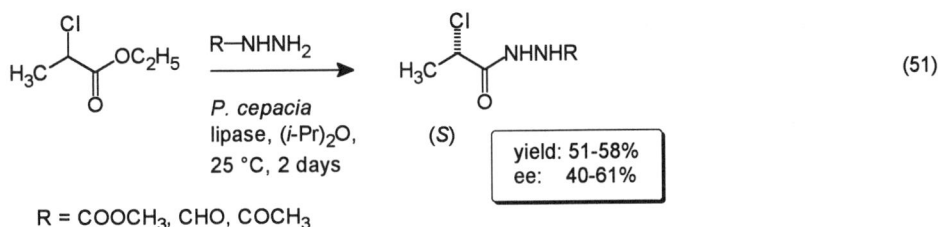

$$R = COOCH_3, CHO, COCH_3$$

Diesters, e.g. dimethyl succinate also undergo hydrazinolysis with hydrazine hydrate or substituted hydrazines in the presence of *Pseudomonas cepacia* lipase[61] to give the corresponding dihydrazide (reaction 52) or monohydrazide (reaction 53), respectively. Interestingly, when the reaction was carried out at 60°C N-amidosuccinimide derivatives were formed (reaction 53).

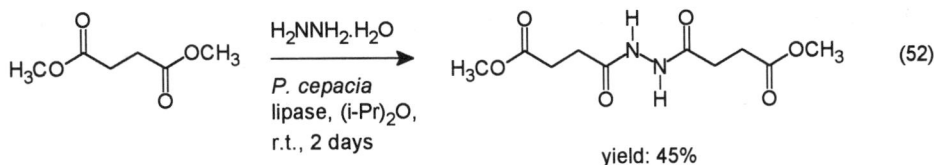

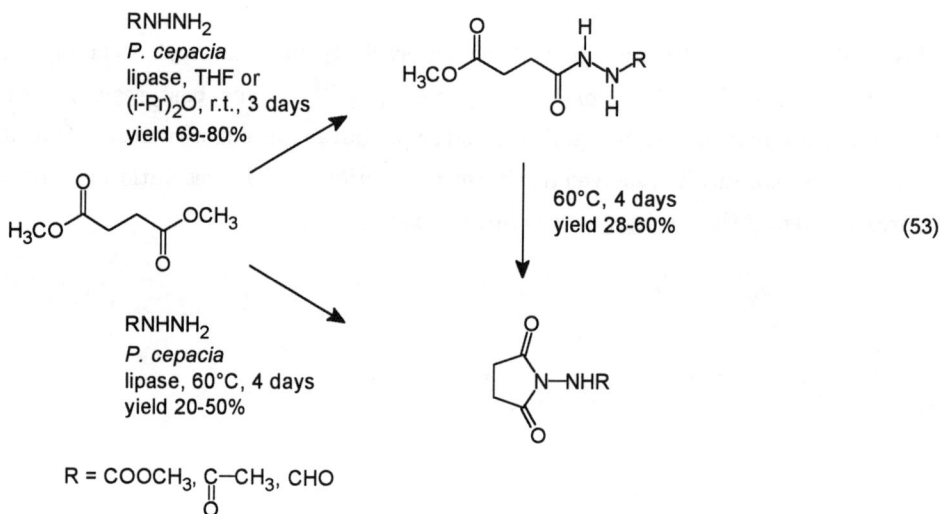

(53)

R = COOCH$_3$, $\underset{\text{O}}{\overset{\text{O}}{\text{C}}}$—CH$_3$, CHO

10. AMINO ACIDS AS ACYL ACCEPTORS: PEPTIDE SYNTHESIS

Peptide bond formation is an important reaction in organic synthesis[68]. Conventional chemical methods have the inherent disadvantage that they require extensive protection and deprotection and racemisation can be a problem. Enzymatic methods, on the other hand, often require minimal protection and proceed under mild conditions. Lipases have been successfully applied to peptide bond formation in organic media under mild conditions. No racemisation is observed and hydrolysis of the peptide product is not a problem under the non-aqueous conditions. An example[69] is the formation of the dipeptide from the N-acylated phenylalanine ester and L-leucine amide illustrated in reaction 54.

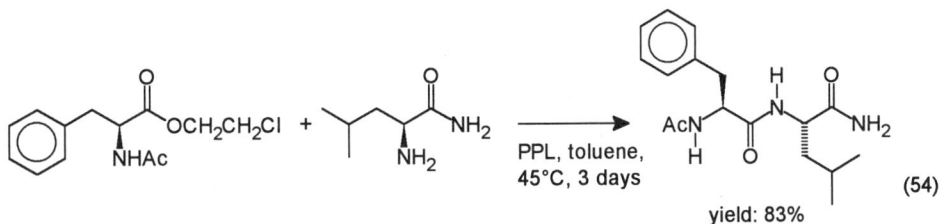

Evidence that the reaction is not catalyzed by protease impurities present in the PPL was provided by the fact that the enzyme preparation did not mediate the hydrolysis of a dipeptide[69]. Other authors[70] in contrast, attributed the high yields of peptide formation observed with N-protected amino acid esters to protease contaminants in the PPL.

Recently, peptide bond formation from Z-amino acid esters and amino acid amides (reaction 55) with PPL in aqueous-water miscible organic solvents was described[71]. However, such results should be treated with caution since bulky substrates, such as Z-Phe-OMe, are generally poor substrates for lipases and PPL preparations are known to be contaminated with proteases.

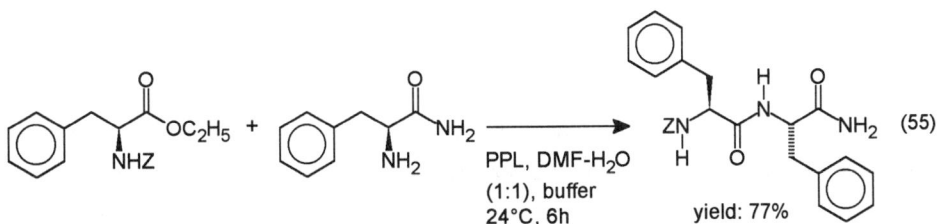

33

11. UNNATURAL ACYL DONORS

Although this review is primarily concerned with lipase-catalyzed reactions involving unnatural acyl acceptors it is worth noting that lipases can also accomodate a variety of acyl donors. The acyl binding site of lipases is shaped like a shallow groove, thus providing an optimum fit for esters of linear carboxylic acids, the natural substrates of lipases. Hence, branched carboxylic acids and their esters generally react much slower than linear ones, up to 10^3 times more slowly if branching is at the α or β carbon atom[72].

Nevertheless, there are many examples of branched carboxylic acids (esters) that are converted at synthetically useful rates. In the *Candida rugosa* lipase-mediated hydrolysis of α-substituted phenylacetic esters[73], the size of the α-substituent had a dramatic effect on the rate (reaction 56).

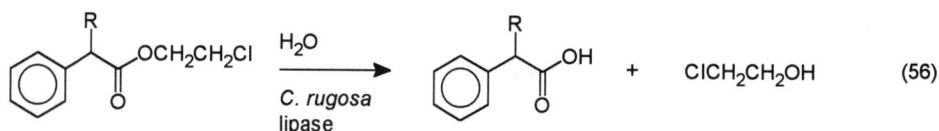

R	rel. rate
methyl	100
ethyl	2.5
cyclohexyl	<<1

Interestingly, different lipases can display opposite preferences for the two enantiomers of chiral carboxylic acids (esters). *Candida rugosa* lipase, for example, generally shows a preference for the enantiomer with the configuration shown in Figure 5[73,74], which could be (*S*) or (*R*) depending on the priority of the α-substituent, X.

Preferred enantiomer:

R^1 = large alkyl or aryl

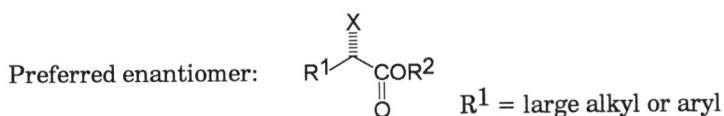

Reaction	Preferred enantiomer	E	Ref.
Esterification		20	74
Hydrolysis			73

X =		
	CH_3	130
	C_2H_5	13
	Br	10
	Cl	4

Figure 5. *Enantiomeric preference of Candida cylindracea lipase.*

Most other lipases, in contrast, show a preference for the opposite enantiomer, i.e. the (R)-enantiomer in the case of α-substituted propionic acids (esters), as illustrated for the transesterification of ketoprofen 2-chloro-ethyl ester[75] in Table 3.

Table 3. *Enantiospecificities of different lipases in the transformation of ketoprofen 2-chloroethyl ester*[75].

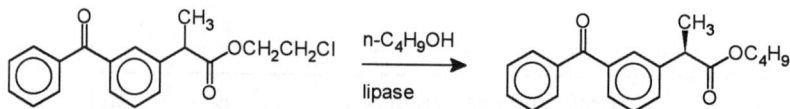

Lipase	E
Amano PS	>100
Amano A. niger	>100
Amano Rh. miehei	>100
Penicillium camemberti	>100
PPL	24
Lipozyme	4

Similarly, *Candida antarctica* lipase shows a preference for the (*R*)-enantiomer in reactions of ibuprofen or its esters with different nucleophiles (Table 4), while *Candida rugosa* lipase is (*S*)-specific.

Table 4. *Enantiospecificities in lipase-catalyzed reactions of ibuprofen (esters) with different nucleophiles, HNu.*

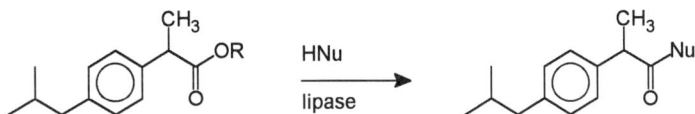

Ref.	Reaction	R	HNu	Lipase	Speci-ficity	conv (%)	ee$_{sub}$	ee$_{prod}$
[76]	esterification	H	1-pentanol	*C. rugosa*	S	49		99
[77]	esterification	H	1-butanol	*C. antarctica*	R	56	86	
[77]	hydrolysis	CH_2CH_2Cl	H_2O	*C. antarctica*	R	63	58	
[77]	ammoniolysis	CH_2CH_2Cl	NH_3	*C. antarctica*	R	56	96	
[74]	hydrolysis	CH_3	H_2O	*C. rugosa*	S	42		95

Whilst the effect of the acyl acceptor on the enantioselection is relatively straightforward[78] the influence of the acyl donor is the result of a multi-step sequence and each step can be medium-dependent[79]. For example, in the *Candida rugosa*- catalyzed esterification of α-aryloxypropionic acids[80] both the magnitude and the sign (*R* or *S*) of the enantiomeric ratio were solvent dependent (Figure 6).

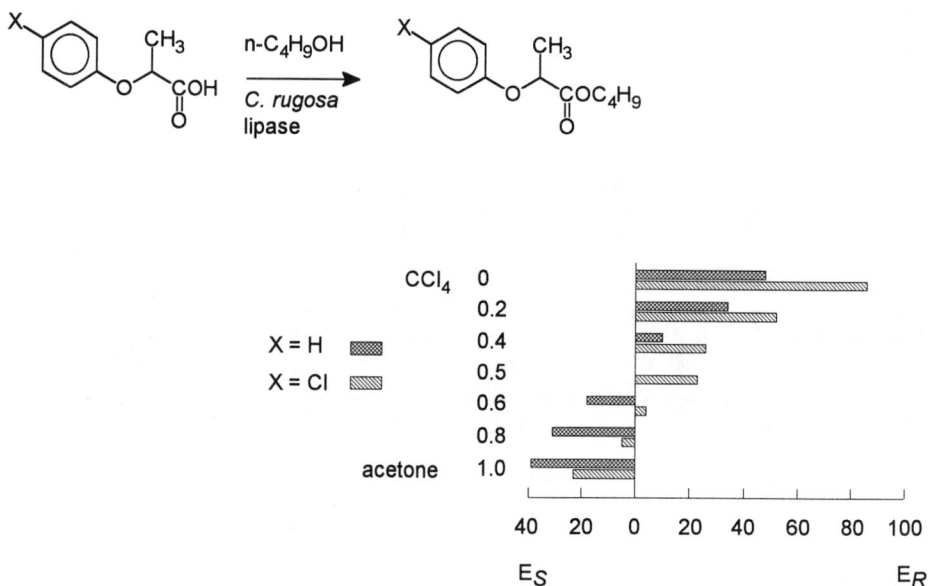

Figure 6. Medium dependence of C. rugosa catalyzed esterification.

11.1. Carbonate esters as acyl donors

Lipases can also accomodate dialkyl and diaryl carbonates as acyl donors[81]. For example, transesterification of diphenyl carbonate with alcohols and phenols[81a] mediated by *Candida rugosa* lipase afforded products derived from substitution of one or both phenoxy groups. Similarly, di-*t*-butyl dicarbonate reacted irreversibly with the weak acyl acceptor acetoxime [81b] as shown in reaction 57.

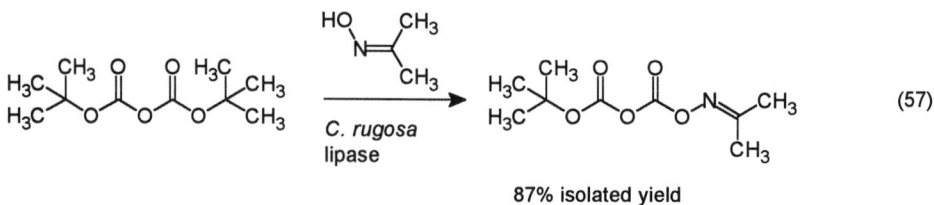

(57)

87% isolated yield

38

Chiral carboxylic acids can be resolved via transesterification of their mixed carboxylic carbonic anhydrides[82] in methyl *t*-butyl ether (MTBE) as solvent (reaction 58). The absolute configuration was not determined.

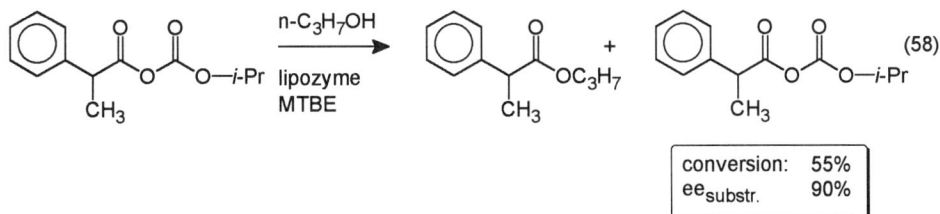

$$\text{(58)}$$

conversion:	55%
ee$_{substr.}$	90%

11.2. Oxazolinones as acyl donors

Oxazolinones, which are easily prepared from α-acylamino carboxylic acids, can also function as acyl donors in lipase-catalyzed reactions[83] as depicted in reaction (59). This constitutes an interesting method for resolving α-amino acids.

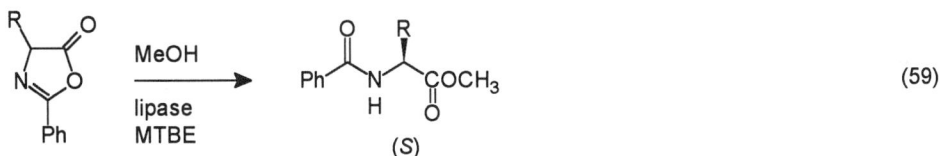

$$\text{(59)}$$

R	Lipase	Yield (%)	ee (%)
4-OH-C$_6$H$_4$	Amano AK	52	75
PhCH$_2$SCH$_2$	Amano K-10	61	90

11.3. Amides as acyl donors

We have already mentioned (see section 10) that lipases catalyze peptide (i.e. amide) bond formation. It has been also shown[84] that amides can function as acyl donors, e.g. (reaction 60).

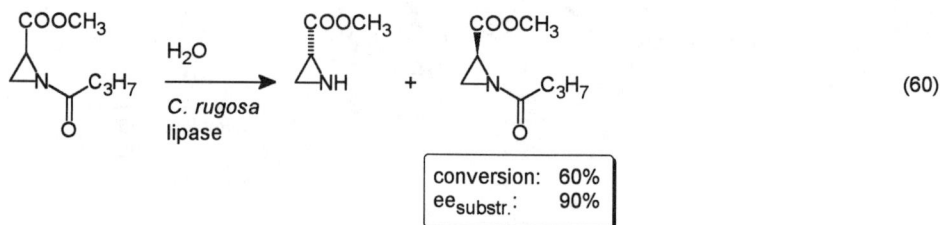

(60)

CONCLUDING REMARKS

Hopefully this review has shown that lipases are remarkably versatile biocatalysts with broad applications in organic synthesis, particularly in the synthesis of pure enantiomers. Their ability to function satisfactorily with a broad range of acyl acceptors and donors as substrates is contrary to conventional wisdom regarding the substrate specificity of enzymes. Coupled with their availibility and relatively good stability this makes lipases eminently suitable for industrial applications. Since many of the reactions discussed have been discovered only in the last five years we confidently expect that their synthetic utility will be extended even further in the near future. One may conclude that once the paradigm of one enzyme - one reaction - one substrate is abandoned the sky is the limit in many biocatalytic transformations.

REFERENCES

1. B. Borgström and H. L. Brockman, (eds.), 'Lipases', Elsevier, Amsterdam, 1984.

2. J. B. Dziezak, *Food Technology*, 114 (1986); D. Stead, *J. Dairy Res.* **53**, 481 (1986); E. W. Seitz, *J. Am. Chem. Soc.* **51**, 12 (1974).

3. H. P. Heldt-Hansen, M. Ishii, S. A. Patkar, T. T. Hansen and P. Eigtved, ACS Symposium series 389, J. R. Whitacker, P. E. Sonnet, (eds.), 'Biocatalysis in Agricultural Biotechnology', American Chemical Society Washington DC, 1989.

4. C. S. Chen and C. J. Sih, *Angew. Chem. Int. Ed. Engl.* **28**, 695 (1989). C. J. Sih, Q. M. Gu, G. Fülling, S. H. Wu and D. R. Reddy, *Dev. Ind. Microbiol.* **29**, 221 (1988).

5. a) G. Kirchner, M. P.Scollar and A. M. Klibanov, *J. Am. Chem. Soc.* **107**, 7072 (1985); b) S. K. Dahod and P. Siuta-Mangano, *Biotechnol. Bioeng.* **30**, 995 (1987).

6. a) M. Kloosterman, V. H. M. Elferink, J. van Iersel, J. H. Roskam, E. M. Meijer, L. A. Hulshof and R. A. Sheldon, *TIBTECH* **6**, 251 (1988); b) L. A. Hulshof, J. H. Roskam (Stamicarbon B. V.) Eur. Pat. Appl. EP 343,714 [C. A.: 113:76603a].

7. F. Björkling, S. E. Godtfredsen and O. Kirk, *J. Chem. Soc., Chem. Commun.*, 934 (1989) .

8. R. H. Valivety, P. J. Halling and A. R. Macrae, *Biochim. et Biophys. Acta* **118**, 218-222 (1992).

9. A. T. J. W. de Goede, W. Benckhuijsen, F. van Rantwijk, L. Maat and H. van Bekkum, *Recl. Trav. Chim. Pays-Bas* **112** , 567 (1993).

10. B. Sjursnes, L. Kvittingen, T. Anthonsen and P. Halling in "Biocatalysis in Non-Conventional Media", J. Tramper (ed), Elsevier Science Publishers B.V. 1992;
    L. Kvittingen, B. Sjursness, T. Anthonsen and P. Halling, *Tetrahedron* **48**, 2793 (1992).

11. A. F. P. Cuperus, S. Th. Bouwer, G. F. H. Kramer and J. T. P. Derksen, *Biocatalysis* **9**, 89 (1994).

12. C-S. Chen, S-H. Wu, G. Girdaukas and C. J. Sih, *J. Am. Chem. Soc.* **109**, 2812 (1987); C-S. Chen and C. J. Sih, *Angew. Chem. Int. Ed. Engl.* **28**, 695 (1989).

13. J. L. L. Rakels, A. J. J. Straathof and J. J. Heijnen, *Enzyme Microb. Technol.* **15**, 1051 (1993).

14. A. L. Margolin, *Enz. Microb. Technol.* **15**, 266 (1993).

15. P. Kalaritis, R. W. Regenye, J. J. Partridge, D. L. Coffen, *J. Org. Chem.* **55**, 812 (1990).

16. G. Braun, P. Braun, D. Kowalczyk and H. Kunz, *Tetrahedron Lett.* **34**, 3111 (1993).

17. P.E. Sonnet and M. W. Baillargeon, *Lipids* **24**, 434 (1989).

18. S. Ueji, R. Fujino, N. Okubo, T. Miyazawa, S. Kurita, M. Kitadani and A. Muromatsu, *Biotechnol. Lett.* **14**, 163 (1992).

19. E. L. A. Macfarlane, F. Rebolledo, S. M. Roberts and N. J. Turner, *Biocatalysis* **5**, 13 (1991).

20. N. Öhrner, M. Martinelle, A. Matson, T. Norin and K. Hult, *Biotechnol. Lett.* **14**, 263 (1992).

21. A. J. M. Janssen, A. J. H. Klunder and B. Zwanenburg, *Tetrahedron* **47**, 7645 (1991).

22. A. M. Klibanov, *Acc. Chem. Res.* **23**, 114 (1990); A. Chogare and J. Sudesh-Kumar, *J. Chem. Soc., Chem. Commun.* 1533 (1989) and 134 (1990).

23. E. Santianello, P. Ferraboschi and P. Grisenti, *Tetrahedron Lett.* **31**, 5657 (1990); Y. F. Wang and C. H. Wong, *J. Org. Chem.* **53**, 3127 (1988); Y. F. Wang, J. J. Lalonde, M. Momongan, D. E. Bergbreiter and C. H. Wong, *J. Am. Chem. Soc.* **110**, 7200 (1988); M. Degueil-Castaing, B. De Joso, S. Droillard and B. Maillaird, *Tetrahedron Lett.* **28**, 953 (1988).

24. B. Berger, A. de Raadt, H. Griengl, W. Hayden, P. Hechtberger, N. Klempier and K. Faber, *Pure Appl. Chem.* **64**, 1085 (1992); K. Faber 'Biotransformations in Organic Chemistry', Springer-Verlag, Berlin Heidelberg, 1992 p 259.

25. D. Bianchi, T. Sugai and H. Ohta, *Agric. Biol. Chem.* **55**, 293 (1991); D. Bianchi, P. Cesti and E. Battistel, *J. Org. Chem.* **53**, 5531 (1988).

26. D. Bianchi and P. Cesti, *J. Org. Chem.* **55**, 5657 (1990).

27. T. Kawamoto, K. Sonomoto, A. Tanaka, *J. Biotechnol.* **18**, 85 (1991).

28. Z. F. Xie, *Tetrahedron Asymm.* **2**, 733 (1991); A. Wickly, E. Schmidt and J. R. Bourne in: J. Tramper (ed.) 'Biocatalysis in Non-Conventional Media', Elseviers Publishers, 1992, pp 577; N. Nakamura, K. Ishihara, A. Ohno, M. Uemura, H. Hayashi, *Tetrahedron Lett.* **31**, 3603 (1990); Y. Yamazaki, K. Hosono, *Agric. Biol. Chem.* **54**, 3357 (1990); Y. Yamazaki, K. Hosono, *Tetrahedron Lett.* **31**, 3895 (1990); T. Itoh, T. Ohta, *Tetrahedron Lett.* **31**, 6407 (1990).

29. G. Guanti, L. Banfi and R. Riva, *Tetrahedron Asymm.* **5**, 9 (1994).

30. F. Theil, H. Schick, M. A. Lapitskaya and K. K. Pivinitsky, *Liebigs Ann. Chem.* 195 (1991).

31. F. Terradas, M. Teston-Henry, P. A. Fitzpatrick and A. M. Klibanov, *J. Am. Chem. Soc.* **115**, 390 (1993).

32. For reviews on enzymatic acylations of sugars see: D. G. Drueckhammer, W. J.Hennen, R. L. Pederson, C. F. Barbas, C. M. Gauteron, T. Krach and C. H. Wong, *Synthesis*, 499 (1991); A. T. J. W. de Goede, M. Woudenberg - van Oosterom and F. van Rantwijk, *Carb. Europe* **10**, 18 (1994).

33. M. Therisod and A. M. Klibanov, *J. Am. Chem. Soc.* **108**, 5638 (1986).

34. N. Khaled, D. Montet, M. Pina and J. Craille, *Biotechnol. Lett.* **13**, 167 (1991); N. Khaled, D. Montet, M. Farinus, M. Pina and J. Craille, *Oleagineux* **47**, 181 (1992).

35. M. Schlotterbeck, S. Lang, V. Wray and F. Wagner, *Biotechnol. Lett.* **15**, 61 (1993).

36. M. Woudenberg - van Oosterom, Delft University of Technology, private communication.

37. S. Riva, J. Chopineau, A. P. G. Kieboom and A. M. Klibanov, *J. Am. Chem. Soc.* **110**, 584 (1988).

38. M. Woudenberg - van Oosterom, manuscript submitted.

39. S. E. Godtfredsen in: L. Copping, R. E. Martin and J. A. Pickett (eds.), in 'Opportunities in Biotransformations', Elsevier, Amsterdam, 1990, p17.

40. A. T. J. W. de Goede, W. Benckhuijsen, F. van Rantwijk, L. Maat and H. van Bekkum, *Recl. Trav. Chim. Pays-Bas* **112**, 567 (1993).

41. B. D. Martin, S. A. Ampofo, R. J. Linhardt and J. S. Dordick, *Macromolecules*

25, 7081 (1992).

42. A. T. J. W. de Goede, M. van Oosterom, M. P. J. van Deurzen, R. A. Sheldon, H. van Bekkum and F. van Rantwijk, *Biocatalysis* **9**, 145 (1994) .

43. F. Björkling, S. E. Godtfredsen and O. Kirk, *J. Chem. Soc., Chem. Commun.* 1301 (1990).

44. F. Björkling, H. Frykman, S. E. Godtfredsen and O. Kirk, *Tetrahedron* **48**, 4587 (1992).

45. M. C. de Zoete, F. van Rantwijk, L. Maat and R. A. Sheldon, *Recl. Trav. Chim. Pays-Bas* **112**, 462 (1993).

46. N. Baba, M. Mimura, J. Hirataka, K. Uchida and J. Oda, *Agric. Biol. Chem.* **52**, 2685 (1988).

47. H. Kitaguchi, P. A. Fitzpatrick. J. E. Huber and A. M. Klibanov, *J. Am. Chem. Soc.* **111**, 3094 (1989).

48. R. G. Bistline, A. Billijk and S. Feairheller, *J. Am. Oil. Chem. Soc.* **68**, 95 (1991).

49. J. Graille, D. Montet, F. Servat, J. Grinaud, G. Renard, P. Galzy, A. Arnaud and L. Marcou, Eur. Pat. Appl., EP 298796 (1987).

50. Z. Djeghaba, H. Deleuze, B. de Jeso, D. Messadi and B. Maillard, *Tetrahedron Lett.* **32**, 761 (1991).

51. B. Tuccio, E. Ferré and L. Comeau, *Tetrahedron Lett.* **32**, 2763 (1991).

52. S. Puertas, R. Brieva, F. Rebolledo and V. Gotor, *Tetrahedron* **49**, 4007 (1993).

53. V. Gotor, R. Brieva, C. Gonzalez and F. Rebolledo, *Tetrahedron* **47**, 9207 (1991).

54. M. J. Garcia, F. Rebolledo and V. Gotor, *Tetrahedron Lett.* **34**, 614 (1993).

55. S. Fernandez, R. Brieva, F. Rebolledo and V. Gotor, *J. Chem. Soc., Perkin Trans I* 2885 (1992).

56. V. Gotor, R. Brieva and F. Rebolledo, *J. Chem. Soc., Chem. Commun.* 957 (1988); in this publication stereochemistry and optical rotations are wrongly assigned; correct assignment is reported in ref. 55 and 57.

57. H. S. Bevinakatti and R. V. Newadkar, *Tetrahedron Asymm.* **1**, 583 (1990); F. Francalanci, P. Cesti, W. Cabri, D. Bianchi, T. Martinengo and M. Foa, *J. Org. Chem.* **52**, 5079 (1987).

58. R. Brieva, F. Rebolledo and V. Gotor, *J. Chem. Soc., Chem. Commun.* 1386 (1990).

59. A. L. Gutman, E. Meyer, X. Yue and C. Abell, *Tetrahedron Lett.* **33**, 3943 (1992).

60. M. C. de Zoete, A. C. Kock-van Dalen, F. van Rantwijk and R. A. Sheldon, *J. Chem. Soc., Chem. Commun.* 1831 (1993).

61. M. C. de Zoete, F. van Rantwijk and R. A. Sheldon (Delft University of Technology), Neth. Appl. nr 93.01574.

62. M. C. de Zoete, A. C. Kock-van Dalen, F. van Rantwijk and R. A. Sheldon, *Biocatalysis* **10**, 307 (1994),.

63. M. C. de Zoete, A. A. Ouwehand, F. van Rantwijk and R. A. Sheldon, this thesis, Chapter 6..

64. C. Astorga, F. Rebolledo and V. Gotor, *Synthesis* 287 (1993).

65. V. Gotor, C. Astorga, F. Rebolledo and E. Menedel, *Indian J. Chem.* **31B**, 906 (1992).

66. V. Gotor in: S. Servi, (ed.), 'Microbial Reagents in Organic Synthesis', Kluwer, Dordrecht, 1992, pp 199-208.

67. C.Astorga, F. Rebolledo and V. Gotor, *Synthesis* 350 (1991).

68. M. Bodansky, 'Principles of Peptide synthesis', Springer-Verlag, Berlin Heidelberg, 1993.

69. A. L. Margolin and A. M. Klibanov, *J. Am. Chem. Soc.* **109**, 3802 (1987).

70. J. Blair West and C. Wong, *Tetrahedron Lett.* **28** 1629 (1987).

71. K. Kawashiro, K. Kaiso, D. Minato, S. Sugiyama and H. Hayashi, *Tetrahedron* **49**, 4541 (1993).

72. P. E. Sonnet and M. W. Baillargeon, *Lipids* **26**, 295 (1991).

73. J. Bojarski, J. Oxelbark, C. Andersson and S. Allenmark, *Chirality* **5**, 154, (1993).

74. P. Berglund, M. Holmquist, E. Hedenström, K. Hult and H.-E. Högberg, *Tetrahedron Asymm.* **4**, 1869 (1993). See also: E. Holmberg, M. Holmquist, E. Hedenström, P. Berglund, T. Norin, H.-E. Högberg and K. Hult, *Appl. Microbiol. Biotechnol.* **35**, 572 (1991); K.-H. Engel, *J. Am. Oil Chem. Soc.* **69**, 146 (1992); Q.-M. Gu and C.J. Sih, *Biocatalysis* **6**, 115 (1992); S.-H. Wu, Z.-W. Guo and C.J. Sih, *J. Am. Chem. Soc.* **112** 1990 (1990).

75. A. Palomer, M. Cabré, J. Ginesta, D. Mauleón and G. Carganico, *Chirality* **5** 320(1993).

76. A. Mustranta, *Appl. Microbiol. Biotechnol.* **38**, 61 (1992).

77. M.C. de Zoete, A.C. Kock-van Dalen, F. van Rantwijk and R.A. Sheldon, *J. Chem. Soc., Chem. Commun.* 1831 (1993).

78. M. Cygler, P. Grochulski, R.J. Kazlauskas, J.D. Schrag, F. Bouthillier, B. Rubin, A.N. Serreqi and A.K. Gupta, *J. Am. Chem. Soc.* **116**, 3180 (1994).

79. K. Hult in: S. Servi (ed.), Microbial Reagents in Organic Synthesis, Kluwer Academic Publishers, Dordrecht, 1992, p 289.

80. S. Ueji, R. Fujino, N. Ōkubo, T. Miyazawa, S. Kurita, M. Kitadani and A. Muromatsu, *Biotechnol. Lett.* **14**, 163 (1992).

81. a: D.A. Abramowicz and C.R. Keese, *Biotechnol. Bioeng.* **33**, 149 (1989); b: M. Menendez and V. Gotor, *Synthesis* 72 (1993).

82. E. Guibé-Jampel and M. Bassir, *Tetrahedron Lett.* **35**, 421 (1994).

83. J.Z. Crich, R. Brieva, P. Marquart, R.-L. Gu, S. Flemming and C.J. Sih, *J. Org. Chem.* **58**, 3252 (1993).

84. M. Bucciarelli, A. Formi, I. Moretti, F. Prati and G. Torre, *Tetrahedron Asymm.* **4**, 903 (1993).

Chapter 2

Lipase-catalyzed formation of peroxyoctanoic acid

Summary

The lipase-catalyzed formation of peroxyoctanoic acid from octanoic acid was studied. Several lipases were tested and the reaction was performed in various solvents. The effect of the reaction temperature was also investigated. It appeared that the initial rate of formation of peroxyoctanoic acid increased with increasing reaction temperature and was highest with lipase SP435 (*C. antarctica* B) as catalyst in acetonitrile as solvent.

INTRODUCTION

Peroxycarboxylic acids[1]

Peroxycarboxylic acids are important peroxidic oxidants in organic synthesis. They have frequently been used in oxidations of e.g. olefins, ketones, amines and sulphides. Most common peroxycarboxylic acids are conveniently prepared by reaction of the corresponding carboxylic acid with hydrogen peroxide catalyzed by a strong acid such as sulfuric acid.

Peroxycarboxylic acids can also be prepared from the corresponding carboxylic acid anhydrides and hydrogen peroxide in the presence of a strong acid catalyst. This procedure is used for the synthesis of monoperoxyphtalic acid, but it is not suitable for the preparation of simple aliphatic peroxycarboxylic acids, because hazardous diacyl peroxides can be formed. Similarly, acyl chlorides can be converted to peroxycarboxylic acids with hydrogen peroxide and pyridine, but again undesirable diacyl peroxides can be formed.

Peroxycarboxylic acids are also formed via the autoxidation of aldehydes. Recently, the formation of peroxycarboxylic acids catalyzed by lipases was described[2]. This subject will be discussed in more detail in a following paragraph.

Physical properties

In "inert" solvents peroxycarboxylic acids form an intramolecular hydrogen bond, resulting in a higher volatility compared to the parent carboxylic acids. The pK values of peroxycarboxylic acids are lower than those for the corresponding carboxylic acids (pK peroxyacetic acid = 8.2, pK acetic acid = 4.75) due to the lack of resonance stabilisation of the percarboxylate anion.

Pure peroxycarboxylic acids derived from short chain carboxylic acids are explosive even at low temperatures. The stability of peroxycarboxylic acids increases with increasing chain length.

Oxidation reactions with peroxycarboxylic acids

1. Epoxidation of olefins

Olefins can be oxidized with peroxycarboxylic acids to give epoxides. The commonly accepted reaction mechanism was first described by Bartlett (Figure 1) which involves a cyclic, non-ionic, three-membered transition state.

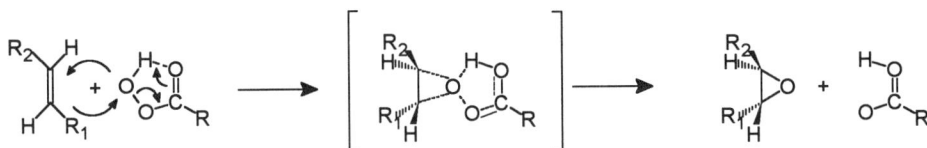

Figure 1. *Reaction mechanism of the epoxidation of olefins by peroxycarboxylic acids.*

The reaction rate of epoxidation is influenced by substituents both in the olefin and the peroxycarboxylic acid. Electron-releasing groups in the olefin accelerate the reaction rate and electron-withdrawing groups retard it, while electron-withdrawing groups in the peroxycarboxylic acid facilitate the reaction, thus indicating the nucleophilic nature of the olefins and the electrophilic nature of the peroxycarboxylic acids in epoxidations.

2. Oxidation of ketones

The oxidation of ketones to esters is known as the Baeyer-Villiger reaction (Figure 2). Oxidation of ketones by peroxycarboxylic acids is slower than epoxidation of olefins and requires strong acid catalysts. The reaction mechanism involves nucleophilic addition of the peroxycarboxylic acid to the protonated carbonyl group with subsequent migration of an electron-rich group to an electrophilic oxygen.
The reaction is accelerated by electron-donating groups in the ketone and by electron withdrawing groups in the peroxycarboxylic acids.

49

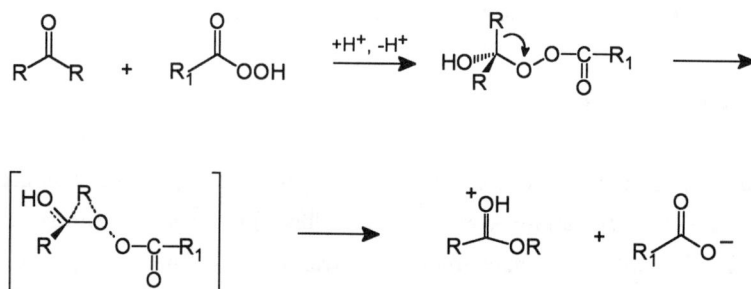

Figure 2. *Reaction mechanism of the Baeyer-Villiger oxidation.*

3. Oxidation of sulfur compounds

Peroxycarboxylic acids oxidize disulfides and thiols to give sulfonic acids. Sulfides are oxidized to sulfoxides, which can subsequently be converted to sulfones. The reaction mechanism of the oxidation is analogous to that of the epoxidation reaction.

4. Oxidation of amines

Amines are oxidized with peroxycarboxylic acids. The products obtained depend on the reaction conditions used. Primary amines are oxidized to nitroso compounds via hydroxylamines; nitroso compounds or their oxime tautomers are further oxidized to nitro compounds under more vigorous conditions. Secondary amines yield nitroxides or nitrones upon oxidation with peroxycarboxylic acids and tertiary amines give the corresponding amine oxides. The reaction mechanism of the oxidation of tertiary amines is analogous to the mechanism of the epoxidation reaction.

THE ENZYMATIC FORMATION OF PEROXYCARBOXYLIC ACID

Workers from Novo reported[2,3] that a lipase from *Candida antarctica* catalyzes the formation of lineair peroxycarboxylic acids from the corresponding carboxylic acid and hydrogen peroxide. An advantage of this procedure is the *in situ* formation of peroxycarboxylic acids under mild reaction conditions, which circumvents the handling problem of an isolated peroxycarboxylic acid. Moreover, stoechiometric amounts of peroxycarboxylic acids are avoided, since the oxidation reaction can be performed with continuous *in situ* regeneration of the peroxycarboxylic acid (Figure 3).

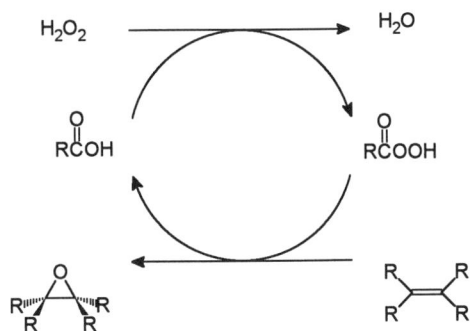

Figure 3. *The use of enzymatically generated peroxycarboxylic acid in oxidation reactions.*

In principle, this technique is applicable to all the oxidation reactions mentioned above. However, the oxidation of amines by peroxycarboxylic acids affords high yields only when an excess of peroxycarboxylic acid is used. Since the enzymatic reaction leads to low concentrations of peroxycarboxylic acid this procedure doesn't seem to be well suited for the oxidation of amines and poor results were observed[3]. We chose the lipase-catalyzed formation of peroxyoctanoic acid from octanoic acid as a model reaction for a systematic investigation of the effects of type of lipase, solvent and reaction temperature.

RESULTS AND DISCUSSION

Standard experimental reaction conditions were as follows:

500 μl (3.15 mmol) octanoic acid and 5 ml solvent were shaken at room temperature with 100 mg lipase as catalyst (except for lipase SP 435 of which also 10 mg was used). 250 μl (5.5 mmol) hydrogen peroxide (60% v/v) was added to start the reaction. The formation of peroxyoctanoic acid was monitored with HPLC.

Several lipase preparations were tested. Five of the lipase preparations displayed catalytic activity: lipase from *Candida antarctica* SP 435, lipase from *Humicola* SP 398 immobilized on Accurel EP100, lipase from *Rhizomucor miehei* immobilized on anionic resin, lipase from *Rhizomucor miehei* immobilized on Accurel EP100 and lipase from Amano P. The lipase from Amano P was an insoluble, probably cross-linked preparation.

The course of the reaction was followed in various solvents (acetonitrile, pentane, tetrahydrofuran, *t*-butyl alcohol and methyl *t*-butyl ether). The influence of the reaction temperature in the lipase SP 435 catalyzed formation of peroxyoctanoic acid was also investigated.

The results of varying the solvent and lipase are presented in figures 4 and 5 (corresponding data are given in the experimental part). Figures 4a-4e compare the formation of peroxyoctanoic acid with various lipases in five different solvents and figures 5a-5e illustrate the solvent effects for the individual enzymes.

Effects of the solvent and the lipase on the formation of peroxyoctanoic acid

The lipase-catalyzed perhydrolysis of octanoic acid is expected to proceed towards an equilibrium on account of the reversibility of the reaction. The position of this equilibrium is independent of the nature and amount of the catalyst. Accordingly, we found that upon reaction of octanoic acid and hydrogen peroxide in MTBE catalyzed by lipase SP 435 the rate of formation of peroxyoctanoic acid subsided after a few hours until a stationary composition was reached eventually. The use of different amounts of enzyme (10 mg and 100 mg of SP435) and addition of fresh lipase did not influence the composition which strongly suggests that an equilibrium had been reached.

The equilibrium is determined by the thermodynamic activities of the reaction components. Since these activities are different in each solvent, the equilibrium is dependent on the solvent. In apolar solvents, such as pentane, water and hydrogen peroxide activity are high (both activities are estimated to be about 0.5), because water forms a separate phase in which hydrogen peroxide is preferably dissolved. In these solvents the addition of an extra portion of hydrogen peroxide after 16 hours of equilibration did not influence the equilibrium yield (pentane: conversion 25%, 84% hydrogen peroxide still present). This suggests that the hydrogen peroxide activity was not influenced. Differences in equilibrium yields in apolar solvents thus depend only on the the differences in the activities of carboxylic acid and peroxycarboxylic acid. In polar solvents, such as acetonitrile, water and hydrogen peroxide activity are low (both activities are estimated to be about 0.05), because water and hydrogen peroxide are miscible in these solvents. Addition of an extra portion of hydrogen peroxide after 16 hours resulted in an increase in the yield of peroxyoctanoic acid. Appearently, the hydrogen peroxide activity increased which resulted in a shift of the equilibrium.

The final concentration of peroxyoctanoic acid strongly depended on the nature of the solvent. Acetonitrile gave the best results in general (up to 35% yield, see Figure 4b) whereas in pentane only 25% peroxyoctanoic acid was formed. This result seems in contradiction to the observation[2] that the highest yields were obtained in non-polar solvents like toluene and hexane.

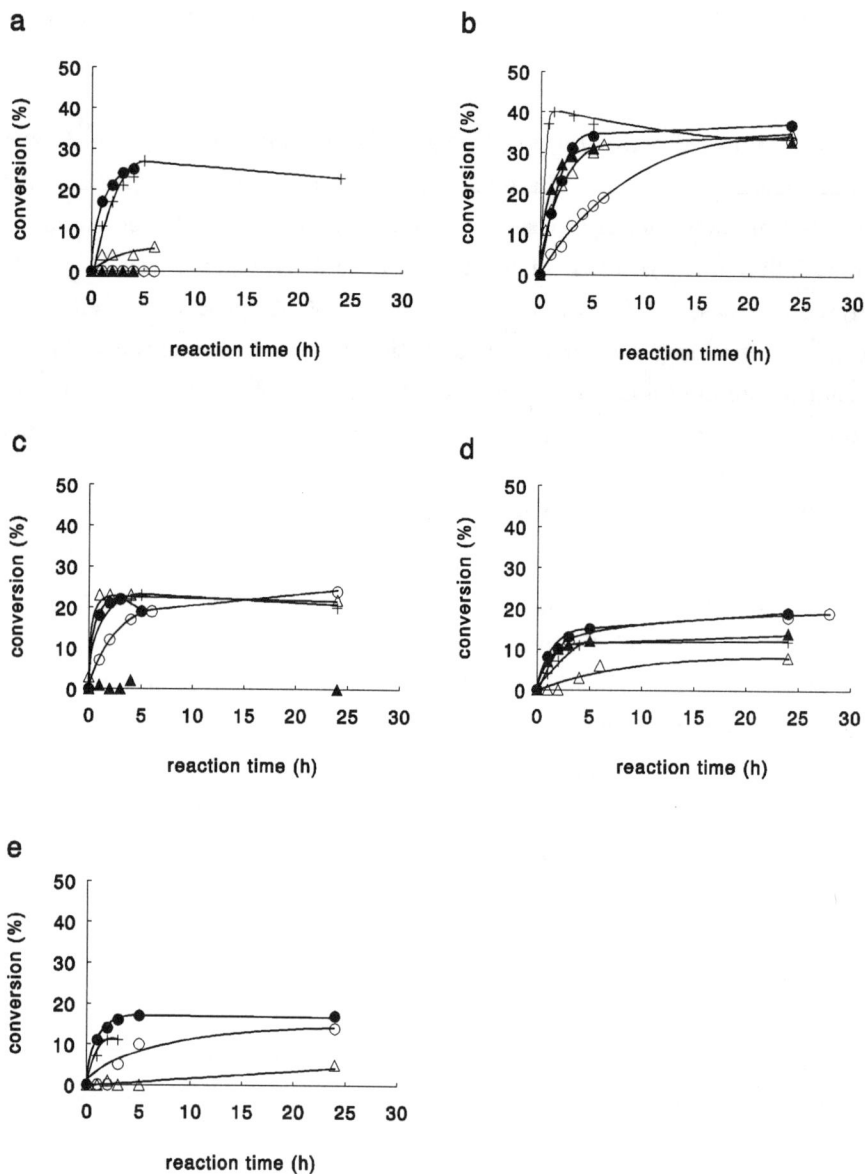

Figure 4. Formation of peroxyoctanoic acid per solvent: a) pentane, b) acetonitrile, c) methyl t-butyl ether, d) tetrahydrofuran, e) t-butyl alcohol; + SP 435, △ SP 398, ○ Amano P, • Rh. miehei on Accurel EP100, ▲ Rh. miehei on anionic resin.

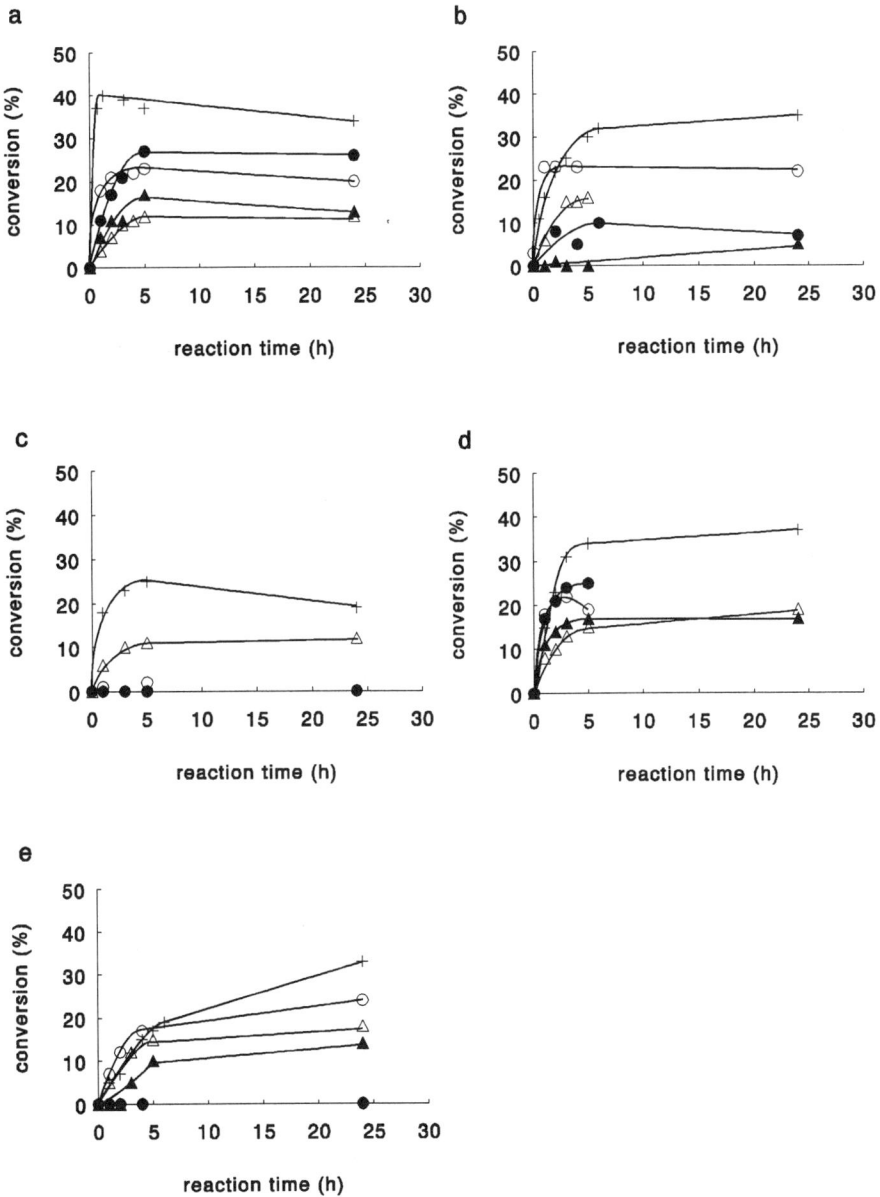

Figure 5. *Formation of peroxyoctanoic acid per lipase, a) SP 435, b) SP 398, c) Rhizomucor miehei on anionic resin, d) Rhizomucor miehei on Accurel EP100, e) Amano P; + acetonitrile, △ THF, ○ MTBE, • pentane, ▲ t-butyl alcohol.*

We ascribe this to the difference in reactant concentrations[†] which would result in different reactant and product activities[‡]. The statement: "lower yields of peroxyoctanoic acid are generally found using water miscible solvents[3]" is not correct for our reaction conditions, however.

In methyl *t*-butyl ether the equilibrium yield is about 22% (Figure 4c). Tetrahydrofuran (Figure 4d) gave inconsistant results: the final yield of peroxyoctanoic acid varied between 5% and 18%. Dissolution or swelling of some of the polymeric supports which consequently will destabilize the enzyme might be involved. In *t*-butyl alcohol (Figure 4d) the equilibrium yield of peroxyoctanoic acid is about 15%.

The initial rate of formation of peroxyoctanoic acid is an important factor to consider when making use of *in situ* generation of peroxycarboxylic acid. Here not only the amount of lipase, but also the nature of the lipase has a large influence, especially when considered together with the solvent system. In this respect SP 435 is clearly superior. Its high perhydrolysis activity allowed for a ten-fold reduction of the amount of enzyme used. Moreover SP435 displayed high stability under the reaction conditions, because it performed well in all solvents (Figure 5a). All other lipases (apart from Amano P) give reasonable production rates in acetonitrile. In pentane, however, SP398, Amano P and *Rhizomucor miehei* on anionic resin are not active, which is presumably caused by deactivation by a high hydrogen peroxide activity. Interestingly, *Rhizomucor miehei* lipase on EP100 performs well in pentane as well as in other solvent systems, whereas *Rh. miehei* on anionic resin is inactive in pentane and MTBE. SP398, which in MTBE has the highest production rate of all catalysts, is nearly inactive in *t*-butyl alcohol. In conclusion, lipase-mediated oxidation reactions can best be performed with SP 435 in acetonitrile.

[†] Our reaction conditions: octanoic acid, 3.15 mmol; H_2O_2, 5.5 mmol; solvent, 5 ml. Conditions stated in ref[2]: octanoic acid, 1.0 mmol; H_2O_2, 2.7 mmol; solvent, 10 ml.

[‡] The same authors reported a yield of 24 % of peroxyoctanoic acid in hexane[3], which is in agreement with our result in pentane (25 %). The initial concentration of octanoic acid was the same in both experiments (3.15 mmol in 5 ml solvent). The amount of hydrogen peroxide was less (4.45 mmol[3]), but the hydrogen peroxide activity will be approximately equal in both systems.
In a patent application by the same authors[4] a yield of 37 % peroxyoctanoic acid is claimed using the same amounts of reactants in xylene. This is probably due to the effect of the solvent on the activity coefficients of octanoic acid and peroxyoctanoic acid.

The effect of the reaction temperature on the formation of peroxyoctanoic acid

The effect of the reaction temperature on the rate of formation of peroxyoctanoic acid was investigated in pentane and acetonitrile with lipase SP 435 as catalyst. The reaction was performed at -13°C, 4°C, 17°C and 37°C in pentane, and in addition to this at -5°C and 60°C in acetonitrile.

From the results it can be seen (figure 6a and figure 6b) that the rate of formation of peroxyoctanoic acid is dependent on the reaction temperature. In pentane the initial reaction rate is highest at 37°C and lowest at -13°C.

In acetonitrile the initial reaction rate at 37°C is as high as at 17°C. At 60°C lipase SP 435 is inactive; probably the enzyme becomes denaturated under these reaction conditions.

The final yield of peroxyoctanoic acid is lower at higher reaction temperature. This is caused by faster degradation of hydrogen peroxide at higher reaction temperature.

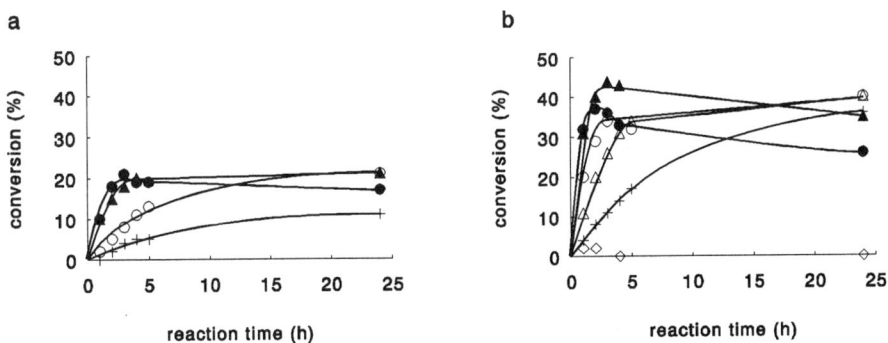

Figure 6. *Effect of the temperature on the formation of peroxyoctanoic acid catalyzed by lipase SP 435 a) in pentane, b) in acetonitrile; +: -13°C, ▵: -5°C, ○: 4°C, ▴: 17°C, ●: 37°C, ◊: 60°C.*

CONCLUSIONS

The lipase-catalyzed formation of peroxyoctanoic acid was succesfully performed in various solvents with five different lipase-preparations. In each system an equilibrium was reached, but the equilibrium yield was dependent on the solvent. High hydrogen peroxide activity, which was obtained in apolar solvents, such as pentane, caused deactivation of the enzyme.

The initial rate of formation of peroxyoctanoic acid was dependent on the reaction temperature and on the solvent. The rate of formation of peroxyoctanoic acid increased with increasing reaction temperature and was highest in acetonitrile as solvent. Since a high initial rate is important for a high product yield when the *in situ* formed peroxyoctanoic acid is used subsequently in oxidation reactions, lipase mediated oxidations can best be carried out in acetonitrile at 17°C or 37°C.

EXPERIMENTAL

Materials and methods

Analytical HPLC was performed using a Waters 590 pump on a reversed-phase column (Nucleosil C18) at ambient temperature with an effluent flow of 1.0 ml/min, with detection on an Erma ERC-7510 RI detector. Eluent: acetonitrile/water : 50/50, 0.02M acetate buffer pH 4.5.

All solvents were of analytical grade.

Accurel EP100 was obtained from Akzo Faser AG.

Lipases used were: *Candida antarctica* SP 435 from Novo Nordisk; *Rhizomucor miehei* on anionic resin from Novo Nordisk; *Humicola* SP 398 from Novo Nordisk, immobilized on Accurel EP100; *Rhizomucor miehei* from Novo Nordisk, immobilized on Accurel EP100 and Amano P from DSM.

Immobilisation procedure

Immobilisation of the lipases on Accurel EP100 was performed according to a published procedure[4].

Standard reaction

500 µl (= 3.14 mmol) octanoic acid and 5 ml solvent were shaken at room temperature with 100 mg lipase as catalyst. 250 µl (= 5.5 mmol) hydrogen peroxide (60%) was added to start the reaction. The formation of peroxyoctanoic acid was monitored with HPLC. Solvents used: acetonitrile, pentane, tetrahydrofuran, *t*-butyl alcohol, methyl *t*-butyl ether. Lipases used were from *Candida antarctica* (SP 435) 10 mg and 100 mg, *Humicola* (SP 398) on Accurel EP100 100 mg, *Rhizomucor miehei* on anionic resin 100 mg, *Rhizomucor miehei* on Accurel EP100 100 mg, and Amano P 100 mg.

Table 1. Formation of peroxyoctanoic acid with 10 mg SP435.

SP 435 (10 mg)	peroxyoctanoic acid formed (%)				
reaction time (h)	acetonitrile	pentane	MTBE	THF	*t*-butyl alcohol
1	37	11	18	4	7
2		17	21	7	11
3	40	21	22	10	11
4			22	11	
5	37	27	23	12	17
24	34	26	20	12	

Table 2. *Formation of peroxyoctanoic acid with 100 mg SP 435.*

SP 435 (100 mg)	peroxyoctanoic acid formed (%)		
reaction time (h)	pentane	MTBE	THF
1	23	21	14
2	24	22	
3	27		15
4		19	
5	22		15
3		22	
24	26	22	15

Table 3. *Formation of peroxycarboxylic acid with SP 398.*

SP 398	peroxyoctanoic acid formed (%)				
reaction time (h)	acetonitrile	pentane	MTBE	THF	*t*-butyl alcohol
1	18		23	6	0
2	22	8	23		0
3	25			15	1
4		5	23	15	0
5	30			16	
6	32	10			0
24	35	7	22		5

Table 4. *Formation of peroxycarboxylic acid with Amano P lipase.*

Amano P	peroxyoctanoic acid formed (%)				
reaction time (h)	acetonitrile	pentane	MTBE	THF	*t*-butyl alcohol
1	5	0	7	5	0
2	7	0	12		0
3	12	0		12	5
4	15	0	17		
5	17	0		15	10
6	19	0	19		
24	33	0	24	18	14

Table 5. *Formation of peroxycarboxylic acid with Rhizomucor miehei lipase on anionic resin.*

Rh. miehei on anionic resin	peroxyoctanoic acid formed (%)			
reaction time (h)	acetonitrile	pentane	MTBE	THF
1	21	0	1	7
3	29	0	0	11
5	31	0	2	12
24	33	0	0	14

Table 6. *Formation of peroxycarboxylic acid with Rhizomucor miehei lipase on Accurel EP100.*

Rh. miehei on Accurel EP100	peroxyoctanoic acid formed (%)				
reaction time (h)	acetonitrile	pentane	MTBE	THF	*t*-butyl alcohol
1	15	17	18	8	11
2	23	21	21	10	14
3	31	24	22	13	16
5	34	25	19	15	17
24	37		19	19	17

Table 7. *Temperature effect on the formation of peroxyoctanoic acid in pentane with lipase SP 435 as catalyst.*

pentane	peroxyoctanoic acid formed (%)			
reaction time (h)	-13°C	4°C	17°C	37°C
1	0	2	10	10
2	2	5	15	18
3	4	8	18	21
4	5	11	20	19
5	5	13		19
24	11	21	21	17

Table 8. *Temperature effect on the formation of peroxyoctanoic acid in acetonitrile with lipase SP 435 as catalyst.*

acetonitrile	peroxyoctanoic acid formed (%)					
reaction time (h)	-13°C	-5°C	4°C	17°C	37°C	60°C
1	4	11	20	31	32	2
2	8	20	29	40	37	2
3	11	26	34	44	36	
4	14	31		43	33	0
5	17	34	32			
24	36	40	40	35	26	0

REFERENCES

1.   Basic principles were taken from the following reviews and references therein: a. The chemistry of peroxides, S. Patai (ed) , Wiley, Chichester, 1983, p 287; b. A. F. Hegarty, in "Comprehensive Organic Chemistry", Barton, Ollis (eds), Pergamon Press, 1979, 1105; c. G. H. Schmid and D. G. Garratt, in "Chemistry of double-bond functional groups", S. Patai (ed), Wiley, London, 1977, 817, ; d. B. Plesnicar, in "Oxidation in Organic Chemistry", W. S. Trahanovsky (ed), Acad. Press, New York, 1978, 210; e. H. O. House in "Modern Synthetic Reactions", W. A. Benjamin (ed), Inc Philippines, 1972, 292.

2.   F. Björkling, S. E. Godtfredsen and O. Kirk, *J. Chem. Soc., Chem. Commun,* 1990, 301.

3.   F. Björkling, H. Frykman, S. E. Godtfredsen, O. Kirk, *Tetrahedron,* 48(22), 4587-4592, 1992.

4.   S. Pedersen and P. Eigtved, WO 90/15868, 27 December 1990, Novo Nordisk A/S.

Chapter 3

Lipase-catalyzed formation of chiral peroxy-carboxylic acids and subsequent epoxidations

Summary

The lipase-catalyzed formation of chiral peroxycarboxylic acids was studied. It appeared that these compounds could be formed in acetonitrile as solvent, but not in pentane. Subsequently, epoxidation reactions with *in situ* formed (chiral) peroxycarboxylic acids were investigated. Although chiral peroxycarboxylic acids were formed and subsequent epoxidation took place, induction of enantioselectivity was not observed.

65

INTRODUCTION

There are only a few examples of asymmetric epoxidation by chiral peroxycarboxylic acids. The most well-known examples are (S)-(+)-peroxycamphoric acid, (S)-(+)-2-phenylperoxypropanoic acid, (R)-(-)-2-(α-naphtyl)peroxypropanoic acid, (R)-(-)-2-cyclohexylperoxypropanoic acid and (S)-(+)-2-methylperoxybutanoic acid[1]. When these were used in the epoxidation of cis- and trans-olefins the epoxides obtained were always optically active although optical yields were rather low (maximum 7.5%).

The use of a chiral "Paynes reagent" (a peroxyimidic acid, formed in situ from the corresponding nitrile and hydrogen peroxide under basic conditions; for structure see below) in asymmetric epoxidation reactions[2], which affords the optically active epoxides in excellent yields and enantiomeric excesses (see table 1), suggests that asymmetric epoxidations with chiral peroxycarboxylic acids should, in principle, be possible.

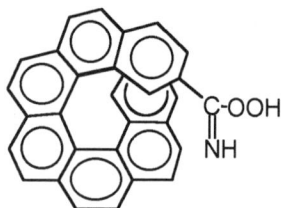

"Paynes reagent" derived from (+)-2-cyano-heptahelicene.

Table 1. Asymmetric epoxidation with a in situ formed peroxyimidic acid derivative of (+)-2-cyanoheptahelicene (a chiral "Paynes reagent").

	yield (%)	ee (%)
α-methylstyrene oxide	84	97.6
trans-stilbene oxide	92	99.8

Asymmetric epoxidation should be favoured by the presence of bulky groups in the chiral peroxycarboxylic acid and the olefin. In this case approach of the peroxycarboxylic acid in such a way that the bulky substituents on the olefin face the least hindered region of the peroxycarboxylic acid, as shown in Figure 1, is favoured.

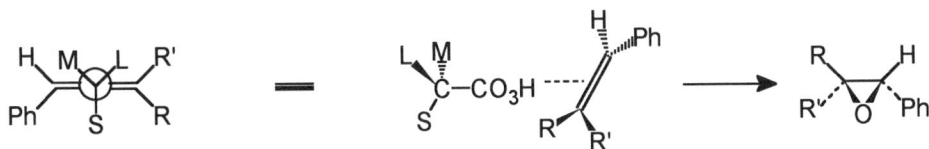

Figure 1. *Asymmetric epoxidation with chiral peroxycarboxylic acids.*

Since lipases catalyze both enantioselective reactions and the formation of peroxyoctanoic acid (Chapter 2), we reasoned that it should be possible to form chiral peroxycarboxylic acids enantioselectively from the parent chiral carboxylic acids. This enzymatic method would yield a mild alternative for the formation of peroxycarboxylic acids and it could be used for the formation of chiral peroxycarboxylic acids substituted with sensitive functional groups. The *in situ* generated chiral peroxycarboxylic acids might subsequently be used for enantioselective epoxidations.

RESULTS AND DISCUSSION

We first had to find a generally applicable method to monitor the formation of chiral peroxycarboxylic acids. In the case of peroxyoctanoic acid this was done with HPLC (Chapter 2), but it was not feasible to monitor the formation of all chiral peroxycarboxylic acids with HPLC, because a new method of analysis would have to be developed for each chiral carboxylic acid. Hence, we used the amount of epoxidation as a measure for the formation of peroxycarboxylic acid (Figure 2).

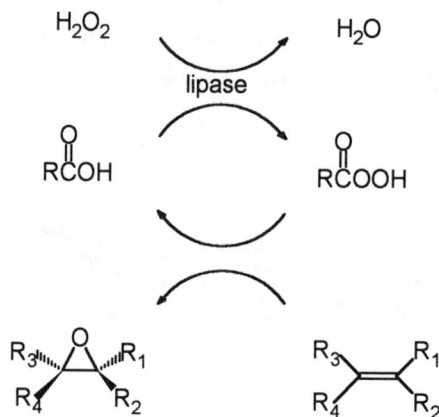

Figure 2. *Formation of peroxycarboxylic acid and subsequent epoxidation.*

The epoxidation of cyclohexene with enzymatically generated peroxyoctanoic acid was chosen as a model reaction[3]. The reaction was catalyzed by lipases from *Candida antarctica*, SP435 (immobilized *Candida antarctica* B) and SP398 (immobilized *Candida antarctica* A and B). The reaction was performed in three reaction media: in acetonitrile, in pentane and in a solvent free system. Hydrogen peroxide was added in small portions to the reaction mixture, because a high concentration of hydrogen peroxide (and thus a high activity of hydrogen peroxide) inactivated the enzyme. Epoxidation was faster in acetonitrile than in pentane (see Figure 3), which is in agreement with our observation that the rate of formation of peroxyoctanoic acid is higher in acetonitrile than in pentane (Chapter 2). In the solvent free reaction the enzyme appeared to be more sensitive to hydrogen peroxide and therefore hydrogen peroxide was added more slowly. Under these reaction conditions a high yield of epoxide was obtained (95% in 24h).

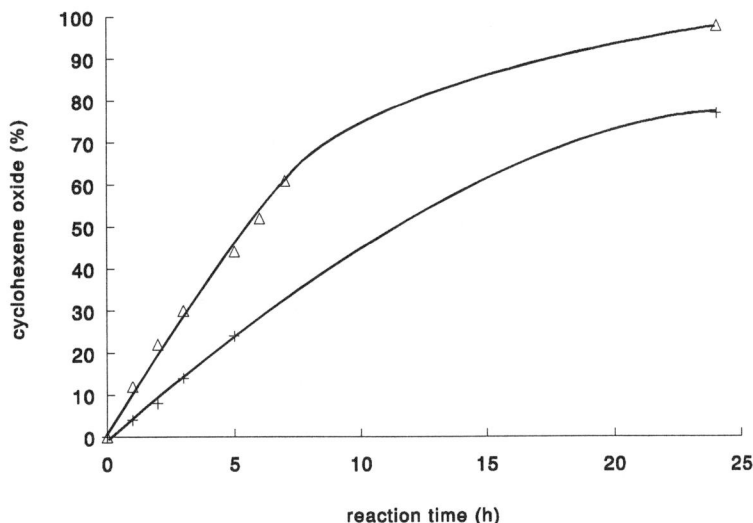

Figure 3: *Epoxidation of cyclohexene in pentane (+) and acetonitrile (△).*

When the formation of peroxycarboxylic acids from α-substituted carboxylic acids was studied, epoxidation of cyclohexene was observed in acetonitrile, but not in pentane or the solvent-free system. Surprisingly, it seemed that no peroxycarboxylic acid was formed in the latter two solvent systems. We checked this by monitoring the formation of 2-methyl-peroxyhexanoic in acetonitrile and in pentane with HPLC. 2-Methyl-peroxyhexanoic acid failed to appear when the reaction was run in pentane, whereas in acetonitrile it was formed in 25% yield after 24 h. This unexpected effect may point to a conformational change in the enzyme at low wateractivity. Therefore, acetonitrile was used as solvent in further experiments.

The rate of epoxidation of cyclohexene in acetonitrile was, as expected, lower with branched carboxylic acids than with linear carboxylic acids (e.g. 2-methyl-hexanoic acid; 34% cyclohexene oxide in 26 h versus octanoic acid; 98% cyclohexene oxide in 24 h).

After having confirmed that peroxycarboxylic acids were formed from chiral carboxylic acids, we investigated the induction of enantioselectivity in epoxidation by chiral peroxycarboxylic acids. *Trans*-2-methylstyrene was chosen as substrate, because of its solubility in acetonitrile and the stability of the corresponding epoxide[4]. In addition to this, we could easily monitor the enantiomeric excess of the epoxide with chiral HPLC. We used both enantiomerically pure carboxylic acids and racemic mixtures in the epoxidation. In all cases, the lipase mediated-epoxidation of *trans*-2-methylstyrene was slower with the chiral carboxylic acids than with octanoic acid. In the cases of lactic acid, 2-chloropropanoic acid, 2-methyl-3-chloropropanoic acid and 2-hydroxy-butanoic acid epoxidation was observed, but only racemic 2-methylstyrene oxide was formed. When "bulkier" carboxylic acids were used such as n-butylmalonic acid and mandelic acid no epoxidation was observed (Table 2).

In conclusion, alkenes can be epoxidized with enzymatically generated peroxycarboxylic acids with high selectivity. Induction of enantioselectivity with chiral peroxycarboxylic acids is, however, not observed. Presumably it is necessary to use "bulkier" carboxylic acids to effect induction of enantioselectivity. These bulky compounds can not be converted into peroxycarboxylic acids by lipases, however, because they are not accepted as substrates. On the other hand, some of these substrates are accepted in lipase-catalyzed ammoniolysis, hydrolysis and alcoholysis e.g. ibuprofen (see chapter 5). A possible explanation for this different behaviour is that the position of the equilibrium in the perhydrolysis reaction is much less favourable than, for example, hydrolysis or alcoholysis.

Table 2. *Epoxidation of trans-2-methylstyrene with peroxycarboxylic acids generated in situ by Candida antarctica lipase-catalyzed reaction with H_2O_2[a].*

Carboxylic acid	Reaction time (h)	Conversion (%)	Amount of epoxide (%)
octanoic acid	48	70	100
(L)-lactic acid	48	44	84
(D,L)-lactic acid	48	47	79
(-)-2-chloro-propanoic acid	48	51	76
(+,-)-2-chloro-propanoic acid	48	39	65
2-methyl-3-chloro-propanoic acid	48	62	92
(R)-2-hydroxy-butanoic acid	30	53	64
(S)-2-hydroxy-butanoic acid	30	53	59
2-hydroxy-1,2,3-propanetricar-boxylic acid	48	0	0
α-hydroxy-phenyl-acetic acid	48	0	0
n-butyl malonic acid	48	0	0
hydroxysuccinic acid	48	0	0
ibuprofen	48	0	0

[a]Conditions: see experimental section.

EXPERIMENTAL

Analytical HPLC was performed using a Waters 590 pump on a reversed-phase column (Nucleosil C18) at ambient temperature and with an effluent flow of 1.0 ml/min, with detection on an Erma ERC-7510 RI detector. Chiral HPLC was performed on a chiral straight-phase column (Baker, Chiralcel OD) at ambient temperature and with an effluent flow of 0.5 ml/min, with detection on a Shimadzu SPD-6A variable wavelength detector at 254 nm. Analytical GC was performed with a Varian Star 3400 on a CP-Sil5 25mx0.32 mm df = 0.12 μm column. All solvents and reagents were reagent grade. Lipases SP382 (immobilized *Candida antarctica* A *and* B) and SP435 (immobilized *Candida antarctica* B) were kindly provided by Novo-Nordisk A/S. (-)-2-chloropropanoic acid, (+,-)-2-chloropropanoic acid and 2-methyl-3-chloropropanoic acid were kindly donated by DSM Andeno.

(*R*)-2-hydroxybutanoic acid and (*S*)-2-hydroxybutanoic acid were prepared enzymatically as previously described.[5]

Epoxidation of cyclohexene with enzymatically generated peroxyoctanoic acid

in pentane and in acetonitrile:
To a solution of cyclohexene (1.5 ml, 14.8 mmol), octanoic acid (300 μl, 1.9 mmol) and octane (200 μl, internal standard) in acetonitrile (5 ml) or pentane (5 ml) was added lipase SP435 (100 mg). Then hydrogen peroxide (60%) was added in 6 portions, each of 150 μl (3.3 mmol), with intervals of one hour. The reaction was monitored with GC. For results see Table 3.

solvent free:
To a solution of octanoic acid (270 μl, mmol) in cyclohexene (1.75 ml, mmol) was added lipase SP382 (175 mg). Then, hydrogen peroxide (60%) was added in 8 portions, each of 100 μl (2.2 mmol), with intervals of one hour. The reaction was monitored with GC. For results see Table 3.

Table 3: *Results of epoxidation of cyclohexene in different reaction media.*

reaction time (h)	cyclohexene oxide (%)		
	pentane	acetonitrile	solvent free
1	4	12	11
2	8	22	20
5	24	44	48
24	77	98	95

Epoxidation of *trans*-2-methylstryrene, with enzymatically generated chiral peroxycarboxylic acids, general procedure.

To a solution of *trans*-2-methylstyrene (200 µl, 8.7 mmol), carboxylic acid (0.4 mmol) diethyleneglycol dibutyl ether (internal standard, 50 µl) in acetonitrile (1 ml) was added lipase SP435 (20 mg). Then hydrogen peroxide (60%) was added in 4 portions each of 20 µl (0.44 mmol) at reaction time t=0, 18 h, 27 h and 45 h. The conversion and enantiomeric excess were monitored with chiral HPLC. For results see Table 2.

REFERENCES

1.   F. Montanari, I. Moretti, G. Torre, *Chem. Commun.*, 135 (1969).

2.   B. Ben Hassine, M. Gorsane, J. Pecher, R. H. Martin, *Bull. Soc. Chim. Belg.* , **95** , 547 (1986).

3.   Enzyme-mediated epoxidation in pentane has been described: F. Bjorkling, S.E. Godtfredsen and O. Kirk, *J. Chem. Soc., Chem. Commun.*, 1301 (1990), F. Bjorkling, H. Frykman, S. E. Godtfredsen and O. Kirk, *Tetrahedron*, **48**, 4587 (1992).

4.   Other substrates tested: *trans*-stilbene and styrene. *Trans*-stilbene did not dissolve in the reaction mixture, epoxidation of styrene resulted in initial formation of styrene oxide, which appeared to polymerize under the reaction conditions (only traces of benzaldehyde and phenylacetaldehyde could be detected).

5.   M.-J. Kim and G. M. Whitesides, *J. Am. Chem. Soc.*, **110**, 2959 (1988); M.-J. Kim and J. Y. Kim, *J. Chem. Soc., Chem. Commun.*, 326 (1991).

Chapter 4

Selective oxidation of penicillin G with hydrogen peroxide and with enzymatically generated peroxyoctanoic acid

Summary

Two convenient procedures for the oxidation of penicillin G into penicillin G-1-(*S*)-sulfoxide have been developed. Both the mild oxidation with enzymatically generated peroxyoctanoic acid and the attractive and simple oxidation with the sole use of hydrogen peroxide are valuable alternatives for the commercial oxidation with 40% peroxyacetic acid.

This Chapter has been published: M. C. de Zoete, F. van Rantwijk, L. Maat and R. A. Sheldon, *Recl. Trav. Chim. Pays-Bas*, 112, 462 (1993).

INTRODUCTION

Penicillin G-1-(S)-sulfoxide has become an important intermediate in the commercial synthesis of cephalosporins since Morin[1] first described its conversion into deacetoxycephalosporin. The sulfoxide is industrially produced by oxidation of penicillin G with 40% peroxyacetic acid and is isolated in a yield of 90-95%[2]. Alternative methods[3] suffer from disadvantages such as long reaction times, expensive reagents, troublesome handling or contamination of the final product by reagents. In all cases only penicillin G-1-(S)-sulfoxide is obtained; This is probably caused by steric interactions.

The lipase-mediated reaction of alkanoic acids with hydrogen peroxide as acyl acceptor[4,5] constitutes a mild procedure for generating peroxycarboxylic acids *in situ*, which may overcome the handling problems of an isolated peroxycarboxylic acid. Complete conversion of penicillin V (**1b**) using this catalytic cycle has been reported[6]. This incited us to apply a similar procedure for the oxidation of the potassium salt of penicillin G (**1a**) (see Figure 1).

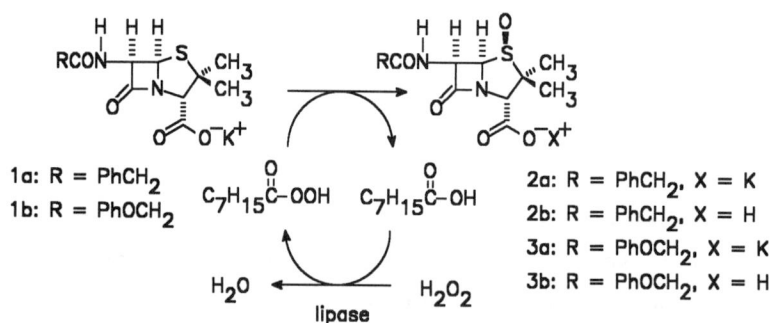

Figure 1. *The oxidation of penicillin G with enzymatically generated peroxyoctanoic acid.*

However, we found that penicillin G sulfoxide (**2**) was obtained in a low yield. During a study to optimize the reaction conditions it became clear that direct oxidation of **1a** by hydrogen peroxide competed with the lipase-mediated process, sometimes eclipsing it, depending on the solvent used.

To our suprise we found that **1a** could be oxidized to **2** with hydrogen peroxide as the only reagent in an aqueous medium. The attractive simplicity of this reaction prompted us to develop a synthetically useful procedure.

As a result, we now report a procedure for direct oxidation of **1a** with hydrogen peroxide in a concentrated aqueous medium which affords **2** in 70% isolated yield, as well as a procedure involving lipase-mediated *in situ* generation of peroxyacid, which affords **2** in 75% yield.

RESULTS AND DISCUSSION

Oxidation of 1a with enzymatically generated peroxyoctanoic acid

The procedure published for the oxidation of **1b**[6] was adapted for the oxidation of **1a** and was studied in toluene[6], *t*-butyl alcohol and acetonitrile. In each solvent, a reaction without enzyme was run in comparison.

In toluene (which formed a two-phase system) and *t*-butyl alcohol low (5%) yields of **2** were obtained, **1a** being mainly converted into degradation products in both the catalyzed and the non-catalyzed reaction.

In acetonitrile the reaction *via* the catalytic pathway predominated (see Table I), which seems in contradiction with the reported lower yield of peroxycarboxylic acid in acetonitrile compared with apolar solvents such as toluene[4,5]. The moderate solubility of **1a** in acetonitrile and the absence of an aqueous phase - and consequently a lower water- and hydrogen peroxide-activity - seem to favour the catalytic cycle. Therefore, further reactions were performed using acetonitrile as the solvent.

The effects of lowering the reaction temperature below ambient and of varying the time between two additions of is shown in Table I. At +4°C and -20°C synthetically useful yields of about 75% were achieved.

Table I. *Lipase-mediated oxidation of penicillin G* (**1a**) *in acetonitrile.*

Reaction temperature (°C)	Reaction time (h)	Time between two additions (min)	Yield with lipase (%)[a]	Yield without lipase (%)[a]
20	6	60	47	16
+4	3.5	30	74	13
+4	6	60	66	-
+4	6	120	59[b]	-
-20	6	60	75	5

[a] determined by HPLC, isolated yield is 80% of yield determined by HPLC.
[b] 13% of starting material was still present in the reaction mixture.

Oxidation of 1a with hydrogen peroxide

During our studies on the lipase-mediated oxidation reaction of **1a** we examined water as solvent. To our surprise an exothermic reaction occurred when hydrogen peroxide (60%) was added at room temperature to a concentrated solution of **1a** in water (1g/ml) before the addition of the lipase. Analysis of the reaction products revealed that **1a** had been converted into **2** in a low yield (5%). When the reaction was performed at 0 and -10°C the yield of **2** increased to 61 and 70% respectively, although the reaction proceeded rather sluggishly at -10°C. Attempts to perform the reaction at -20°C failed because the reaction mixture solidified.

We also attempted to run a reaction by adding 60% hydrogen peroxide to solid **1a** at 0°C, because it seemed that high concentrations of the reactants enhanced the yield. However, a violent exothermic reaction took place and no product (**2**) could be detected.

The yield of **2** at -10°C (70%) is comparable with the yield of the lipase-mediated reaction, but lower than the yield of the industrially applied oxidation with peracetic acid (90-95% isolated yield). Although hydrogen peroxide (60%) entails certain risks, the handling of peracetic acid, which is used in industrial processes at the moment, is more dangerous[7]. In our view, the procedure of the direct oxidation is a valuable alternative, because of its simplicity and the fact that it forms only innocuous water as the coproduct.

EXPERIMENTAL

^{1}H and ^{13}C NMR spectra were recorded in CDCl$_3$ with a few drops of DMSO and with TMS as internal standard using a Varian VXR-400S spectrometer. Analytical HPLC was performed using a Waters 590 pump on a reversed-phase column (8 x 100 mm, Nucleosil C$_{18}$, 10 μm) at ambient temperature and an eluent flow of 1.0ml/min, with detection on an ERMA RI-detector ERC-7510 and a Shimadzu SPD-2A variable wave length detector at 220 nm. Eluent: Acetonitrile (85%), H$_2$O (15%), 0.02M Na$_2$HPO$_4$/H$_3$PO$_4$ at pH6. Melting points are uncorrected and were measured on a Büchi 510 melting point apparatus. The lipase used was an immobilized preparation of *Candida antarctica* lipase SP 435.

Oxidation of 1a with enzymatically generated peroxyoctanoic acid

Hydrogen peroxide (60%) (0.1 ml, 2.2 mmol) was added to a solution of octanoic acid (0.22g, 1.5 mmol) in 5 ml of solvent (toluene, *t*-butyl alcohol or acetonitrile) containing 50 mg of immobilized *Candida antarctica* lipase at room temperature. The reaction mixture was shaken for 2 h at ambient temperature and then cooled to the required reaction temperature. At +4°C **1a** (690 mg, 1.9 mmol in all) was added in three equal portions with intervals of 30, 60 or 120 min (as indicated in Table 1), followed by addition of hydrogen peroxide (60%) (0.2 ml, 4.4 mmol in all) in two equal portions together with the first and second portion of **1a**.
At room temperature and -20°C **1a** (690 mg, 1.9 mmol in all) was added in three equal portions every 60 min and followed by addition of H$_2$O$_2$ (60%) (0.3 ml, 6.6 mmol in all) simultaneously with **1a**. Yields were determined with HPLC.
After the required reaction time (as indicated in Table 1) 10 ml of H$_2$O was added.

The aqueous layer was washed with CH_2Cl_2 (3 x 5 ml) to remove the octanoic acid. The organic layer was discarded and to the aqueous layer 10 ml of CH_2Cl_2 was added followed by acidification of the aqueous layer to pH 1.5 with 1M H_2SO_4. The aqueous layer was extracted with CH_2Cl_2 (3 x 10 ml). The combined organic layers were washed with brine (2 x 10 ml) and dried over $MgSO_4$. After evaporation of the solvent *in vacuo* **2b** was obtained in 80% of the yield determined with HPLC.

^{1}H NMR (400 MHz):δ 1.28 [s, 3H, C$\underline{H}_3$], 1.71 [s, 3H, C$\underline{H}_3$], 3.59 [s, 2H, C$\underline{H}_2$], 4.54 [s, 1H, C$\underline{H}$COOH], 5.07 [d, 1H, J=4.7, C$\underline{H}$SO], 5.96 [dd, 1H, J=4.7, J=10.2, C$\underline{H}$NH], 7.30 [m, 5H, $C_6\underline{H}_5$], 7.40 [d, 1H, J=10.2, N$\underline{H}$], 7.50 [s, 1H, COO$\underline{H}$].

^{13}C NMR (400 MHz): δ 18.46, 19.24, 43.13, 56.05, 66.32, 74.98, 76.38, 127.18, 128.72, 129.18, 134.11, 169.64, 170.54, 173.70.

Mp: 141-146°C (dec) Ref.3: 142-143°C.

The product obtained was in all aspects identical with pure **2b**.

Oxidation of 1a with hydrogen peroxide

At 0°C, hydrogen peroxide (60%) (2.1 ml, 45.9 mmol) was added to a solution of **1a** (3.72 g, 10 mmol) in 3.7 ml of H_2O. The reaction mixture was stirred at 0°C for 6 h until completion. (The reaction was monitored with HPLC). Then, 25 ml of H_2O and 25 ml of CH_2Cl_2 were added at 0°C. The mixture was acidified to pH 1.5 with 1M H_2SO_4. The aqueous layer was extracted with CH_2Cl_2 (3 x 10 ml). The combined organic layers were washed with H_2O (3 x 10 ml) and dried over $MgSO_4$. Evaporation of the solvent *in vacuo* yielded **2b** (2.2 g, 6.1 mmol, 61%).

At -10°C, 3.7 ml of H_2O and H_2O_2 (60%) (2.1 ml, 4.6 mmol) were cooled to -10°C and **1a** (3.72 g, 10 mmol) was added. The reaction was stirred at -10°C for 17 h. The reaction was monitored with HPLC. The reaction product **2b** (2.5 g, 7.0 mmol, 70%) was isolated by the procedure described for the reaction at 0°. The product obtained was in all aspects identical with the compound described above.

ACKNOWLEDGEMENTS

We thank NOVO-Nordisk A/S, Denmark, for the gift of immobilized *Candida antarctica* lipase. The potassium salt of penicillin G was kindly donated by Gist-brocades N.V., The Netherlands.

REFERENCES

1. R. B. Morin, B. G. Jackson, R. A. Mueller, E. R. Lavagnino, W. B. Scanlon and S. L. Andrews, *J. Am. Chem. Soc.* **85**, 1896 (1963).
2. Personal communication of Dr. A. P. G. Kieboom, Gist-brocades N.V., Delft, The Netherlands.
3. A. W. Chow, N. M. Hall and J. R. E. Hoover, *J. Org. Chem.* **27**, 1381 (1962), D. O. Spry, *J. Org. Chem.* **37**, 793 (1972), R. R. Chauvette, P. A. Pennington, C. W. Ryan, R. D. G. Cooper, F. L. Jóse, I. G. Wright, E. M. van Heyningen, G. W. Huffman, *J. Org. Chem.* **36**, 1259 (1971), C.R. Harrison and P. Hodge, *J. Chem. Soc., Perkin I*, 2252 (1976), A. Palomo Coll, (Gema S.A.), Span. ES 509,855 [C.A. **99**, 158134a (1983)],Y. Yoshioka and M. Tanaka, (Shionogi and Co., Ltd.), Jpn. Kokai Tokkyo Koho JP 61 50,986 [C.A. **105**, 114836u (1986)], A. Mangia, *Synthesis* 361 (1978), E. H. Flynn, "Cephalosporins and Penicillins, Chemistry and Biology", Academic Press, New York, p667 (1972), E. Guddal, P. Mørch and L. Tybring, *Tetrahedron Lett.* **9**, 381 (1962), G. V. Kaiser, R. D. G. Cooper, R. E. Koehler, C. F. Murphy, J. A. Webber, I. G. Wright and E. M. Van Heyningen , *J. Org. Chem.* **35**, 2430 (1970), P. G. J. Nieuwenhuis, (AKZO N.V.), PCT Int. Appl. WO 91/05788 [C.A. **115**, 28986t (1991)].
4. F. Björkling, S. E. Godtfredsen and O. Kirk, *J. Chem. Soc., Chem. Commun.* 1301, (1990).
5. F. Björkling, H. Frykman, S.E. Godtfredsen and O. Kirk, *Tetrahedron* **48**, 4587 (1992).
6. O. Kirk, F. Björkling and S. E. Godtfredsen, (NOVO-Nordisk A/S), PCT Int. Appl. WO 91/04333 [C.A. **115**, 181544s (1991)].
7. G. Hommel, Handbuch der Gefährlichen Guter, Springer-Verlag, Berlin Heidelberg (1987).

Chapter 5

A new enzymatic reaction: Ammoniolysis of carboxylic esters.

Summary

A new enzymatic reaction of carboxylic esters and ammonia (ammoniolysis) was studied. This reaction provides a synthetically useful and mild alternative for the synthesis of amides. Several lipases and one esterase acted as catalyst. Ammoniolysis of esters of chiral carboxylic acids gave higher ee values than hydrolysis under comparable reaction conditions. Furthermore, consecutive enzymatic esterification and ammoniolysis provided a convenient one-pot synthesis of carboxylic amides from carboxylic acids.

Parts of this chapter have been published:

Marian C. de Zoete, Alida C. Kock-van Dalen, Fred van Rantwijk and Roger A. Sheldon, *J. Chem. Soc., Chem. Commun*, 1831 (1993);

M. C. de Zoete, A. C. Kock-van Dalen, F. van Rantwijk and R. A. Sheldon, *Biocatalysis*, **10**, 307 (1994).

INTRODUCTION

In the last decade the use of enzymes in synthetic chemistry has been growing continuously. It is now well established that hydrolytic enzymes, such as lipases, esterases and proteases are stable in organic solvents and can be used for enantioselective hydrolysis, esterification and transesterification[1] reactions. This has found application in the resolution of chiral alcohols and carboxylic esters.

Recent research has shown that hydrolytic enzymes can also be used for the acylation of unnatural acyl acceptors like hydrogen peroxide[2], alkylamines[3] and hydrazines[4] according to the general scheme 1.

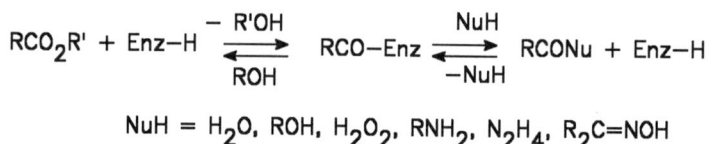

$$RCO_2R' + Enz-H \underset{ROH}{\overset{-\ R'OH}{\rightleftharpoons}} RCO-Enz \underset{-NuH}{\overset{NuH}{\rightleftharpoons}} RCONu + Enz-H$$

$$NuH = H_2O, ROH, H_2O_2, RNH_2, N_2H_4, R_2C{=}NOH$$

Scheme 1. *General lipase-catalyzed reaction.*

With respect to amidation reactions catalyzed by hydrolases, these have been applied in the synthesis of peptides from protected D- or L-amino acids[5] and in the aminolysis for the resolution of chiral carboxylic acids and chiral amines[6].

Surprisingly, the use of ammonia as acyl acceptor has not been reported. The reaction of carboxylic esters and ammonia (ammoniolysis) is efficiently catalyzed by several lipases and one esterase. This reaction provides a mild procedure for the preparation of carboxylic amides as well as an attractive method for the resolution of chiral acids and alcohols.

Furthermore, we developed a one-pot procedure for the conversion of carboxylic acids into their amides consisting of consecutive enzymatic esterification and ammoniolysis. This procedure is expected to give high enantioselectivity when applied to chiral carboxylic acids.

RESULTS AND DISCUSSION

Screening of enzyme preparations

About twenty lipase, esterase and protease preparations (see experimental for details) were screened as catalysts of the reaction between ethyl octanoate and ammonia in *t*-butyl alcohol. Of these, seven lipase-preparations and one esterase displayed ammoniolytic activity (see Table 1) when ammonia saturated *t*-butyl alcohol was used. The *C. antarctica* lipases SP435 and SP525 gave the best results and lipase SP435 was therefore used as catalyst in further experiments. The other enzyme preparations did not show ammoniolytic activity under these reaction conditions. However, when one equivalent of ammonia was used (with respect to the amount of ethyl octanoate), the lipases from Amano PS, *Rhizomucor miehei*, porcine pancreas and SP526 also displayed activity. In the cases of lipase SP435, SP398 and SP523 the yields of octanamide were lower than with ammonia saturated *t*-butyl alcohol. The use of *in situ* generated ammonia from ammonium chloride and triethylamine, which use was described for ammoniolysis of amino acid esters catalyzed by the protease "alcalase", was not suitable for most enzyme preparations. Most of them lost activity, probably due to the basicity of triethylamine.

Table 1. *Ammoniolytic activity of tested enzyme preparations at different ammonia concentrations.*

Enzyme preparation	Conversion[a] of ethyl octanoate (%) in 17 h, unless described otherwise		
	NH$_3$-saturated	Equimolar NH$_3$	NH$_3$ from NH$_4$Cl and triethylamine
C. antarctica lipase SP435	95 in 24 h	43	47
C. antarctica lipase SP525	100 in 24 h	55 in 24 h	87 in 24 h
Pseudomonas alcaligenes	95 in 24 h	53 in 24 h	26 in 24 h
Humicola lipase SP398	60 in 24 h	23	6
Humicola lipase SP523	85 in 24 h	38	13
Pseudomonas lipoprotein lipase	15 in 24 h	34	11
Rhizopus arrhizus lipase	10 in 72 h	8	1
C. cylindracea cholesterol esterase	5 in 72 h	0	0
Amano PS lipase	0	7	1
Rhizomucor miehei lipase	0	20	1
porcine pancreas lipase	0	12	4
SP526	0	2	0

[a] The conversion was determined by HPLC

Ammoniolysis of fatty acid esters

The ammoniolysis of ethyl octanoate and ethyl butyrate (see Figure 1) with lipase SP435 was studied under several reaction conditions. The best results were obtained by using ammonia-saturated *t*-butyl alcohol which was prepared by bubbling gaseous ammonia through *t*-butyl alcohol during one hour at room temperature. For both substrates a reaction without enzyme was run in comparison, but no amide was formed without enzyme.

Under these mild reaction conditions complete conversion of ethyl octanoate and ethyl butyrate was obtained in 24 h. According to HPLC analysis 95% of amide and 5% of acid (due to traces of water) were formed.

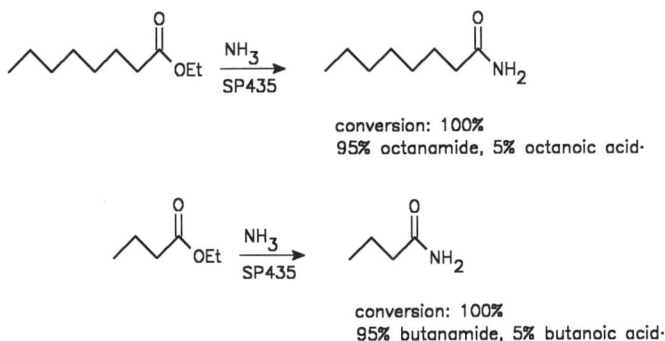

conversion: 100%
95% octanamide, 5% octanoic acid·

conversion: 100%
95% butanamide, 5% butanoic acid·

Figure 1. *Ammoniolysis of fatty acid esters.*

Fatty acid amides ($\geq$ C_{12}) are important industrial products, which are manufactured at approximately 50×10^3 t/y worldwide. Their manufacture involves a high temperature conversion (200°C, 350-700 kPa) followed by distillation. We found that trilaurin, which is, *inter alia*, a natural substrate of lipases, is smoothly converted into laurinamide by lipase-catalyzed ammoniolysis (isolated yield 97%). Due to the lower reaction temperature and the selectivity of the reaction a purer product should be obtained, especially in the case of thermolabile unsaturated compounds like oleic and eruic amide. Therefore, the lipase-catalyzed ammoniolysis of triglycerides provides an attractive alternative method for the synthesis of fatty amides.

Ammoniolysis of esters of chiral carboxylic acids

Although the above-described enzymatic ammoniolysis is of synthetic value as such its utility would be greatly enhanced if it exhibited enantioselectivities superior to conventional hydrolytic processes. In order to compare the enantioselectivity of ammoniolysis versus hydrolysis and alcoholysis several esters of α-substituted carboxylic acids were subjected to ammoniolysis with lipase SP435 as the catalyst in *t*-butyl alcohol. Ethyl 2-chloropropionate, ethyl lactate, ethyl 2-hydroxyhexanoate and ethyl 2-methylbutyrate were converted into the corresponding amides with low to moderate ee.

The ammoniolysis of ibuprofen 2-chloroethyl ester, however, was highly enantioselective. At 56% conversion (48 h) the ee of the remaining ester ((S)-(+)-enantiomer) was 96%. (see Figure 2).

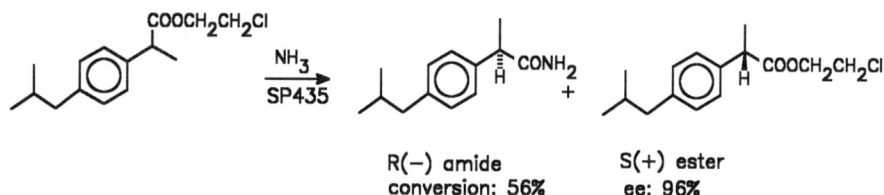

Figure 2. *The ammoniolysis of ibuprofen 2-chloroethyl ester.*

Ester-ammoniolysis is an irreversible reaction (experiments concerning enzymatic conversion of the amide into the corresponding ester revealed that amides were not accepted as acyl donors by lipases). Therefore it is allowed to use the following equations for the calculation of the enantiomeric ratio (E) of the ammoniolysis:

For the product:

$$E = \frac{\ln[1-c(1+ee_p)]}{\ln[1-c(1-ee_p)]}$$

For the substrate:

$$E = \frac{\ln[(1-c)(1-ee_s)]}{[\ln(1-c)(1+ee_s)]}$$

The enantiomeric ratios (E) of the ammoniolysis and the hydrolysis reaction were compared for ibuprofen 2-chloroethyl ester and ethyl 2-chloropropanoate. Ammoniolysis appeared to be an order of magnitude more enantioselective than hydrolysis; for ibuprofen 2-chloroethyl ester ammoniolysis was also more enantioselective than esterification and transesterification (see Table 2).

Table 2. *The enantioselectivity of ammoniolysis versus hydrolysis.*

substrate	E ammoniolysis	E hydrolysis	E esterification	E transesterification
ethyl 2-chloropropanoate	6	3		
ibuprofen 2-chloroethyl ester	28	4	8	16

Many resolutions are performed by hydrolysis of esters of chiral carboxylic acids. Ammoniolysis, however, is a better alternative since it is more enantioselective than hydrolysis.

Ammoniolysis of esters of chiral alcohols

We reasoned that ammoniolysis would also be a useful method for the resolution of chiral alcohols via their esters. Hence, we performed the ammoniolysis of α-methylbenzyl n-butyrate (see Figure 3) in *t*-butyl alcohol as solvent and lipase SP435 as catalyst. (*R*)-(+)-α-methylbenzyl alcohol was formed with high ee.
The enantioselective ammoniolysis of α-methylbenzyl n-butyrate was compared with the transesterification by ethanol and with the hydrolysis. All reactions are highly enantioselective (ammoniolysis: conversion 45%, ee 98%, E > 100; transesterification: conversion 46%, ee 96%, E > 100; hydrolysis: conversion 47%, ee 96%, E > 100).

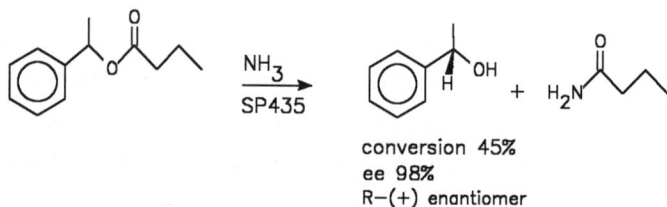

conversion 45%
ee 98%
R–(+) enantiomer

Figure 3. *The ammoniolysis of α-methylbenzyl n-butyrate*

Thus, the ammoniolysis is not only a very enantioselective reaction for the resolution of chiral carboxylic acids, but also for the resolution of chiral alcohols.

One-pot synthesis of octanamide from octanoic acid

Since lipases also catalyze esterifications we reasoned that it should be possible to carry out the conversion of a carboxylic acid to the corresponding amide via *in situ* formation of the ester. (Although direct aminolysis from carboxylic acids into N-octyl-alkyl amide is reported[7] direct ammoniolysis of octanoic acid not feasible because the ammonium salt of the carboxylic acid is not a suitable substrate). This indeed proved to be the case. In this procedure octanoic acid (0.5 ml) was esterified with ethanol and lipase SP435 as catalyst at 40°C. Ethanol was used both as reactant and as solvent. After 24 h 95% ethyl octanoate was formed. The enzyme was filtered off and the ethanol evaporated *in vacuo*. Then, *t*-butyl alcohol and lipase SP435 were added. Ammonia was bubbled through the solution and after 24 h 80% of octanamide and 10% of octanoic acid were formed, the remaining (10%) being non-reacted ethyl octanoate. The formation of octanoic acid is probably due to traces of water, which are formed in the esterification and are not removed by evaporation *in vacuo*.

Because a real "one-pot synthesis" of octanamide would be very attractive, we attempted to perform the esterification and the ammoniolysis consecutively in n--butyl alcohol. Although n-butyl alcohol will compete with ammonia as nucleophile in the ammoniolysis, complete conversion to octanamide can be reached, in principle, because ammoniolysis is an irreversible reaction while esterification is reversible.

The esterification of octanoic acid (5 ml) was performed with n-butyl alcohol as reactant as well as solvent at deminished pressure, under reflux at 50°C, with lipase SP435 as catalyst. The refluxing n-butyl alcohol was dried over molsieves, to exclude the hydrolysis side reaction. Butyl octanoate was readily formed (98% in 24 h). However, after saturation of the n-butyl alcohol with ammonia to perform the ammoniolysis, formation of octanamide was not observed.

This unexpected behaviour was studied by performing the ammoniolysis of butyl octanoate in mixtures of *t*-butyl alcohol and n-butyl alcohol. It appeared that n--butyl alcohol strongly inhibited lipase SP435 in the ammoniolysis reaction, because the conversion of butyl octanoate dramatically decreased in the presence of n-butyl alcohol (Table 3).

To study whether this behaviour was an intrinsic feature of all lipases or only of lipase SP435, a number of other lipase-preparations were tested in the ammoniolysis of butyl octanoate in mixtures of *t*-butyl alcohol and n-butyl alcohol. All lipase-preparations showed a decrease in conversion after 24 h at an increasing amount of n-butyl alcohol, which was to be expected because n-butyl alcohol will compete with NH_3 as acyl acceptor. However, from the data in Table 3 it is evident that lipase SP435 is more strongly inhibited by n-butyl alcohol (9% conversion in 24 h at 50/50 *t*-BuOH/n-BuOH) than lipase SP525 (42% conversion in 24 h at 50/50 *t*-BuOH/n-BuOH). Since lipase SP525 is the same lipase as used in the lipase preparation SP435, but immobilized on Accurel EP100 instead of on polyacrylate, the inhibition of lipase SP435 in the ammoniolysis by n-butyl alcohol seems merely an effect of the support.

Table 3. *Ammoniolytic activity of lipase-preparations in mixtures of t-butyl alcohol and n-butyl alcohol.*

lipase-preparation	t-BuOH/ n-BuOH	conversion (%) in 24 h
SP435	100/0	77
	50/50	9
	0/100	5
SP525 on EP100	100/0	99
	50/50	42
	0/100	13
SP523 on EP100	100/0	41
	50/50	27
	0/100	8
P. alcaligenes on EP100	100/0	80
	50/50	23
	0/100	11
Rh. miehei on anionic resin, 1 eq NH$_3$	100/0	15
	50/50	0
	0/100	0

To avoid the inhibition of SP435 by n-butyl alcohol in the ammoniolysis the esterification was performed with one equivalent of n-butyl alcohol in *t*-amyl alcohol as solvent. The reaction was performed at deminished pressure and under reflux over molecular sieves to absorb the water side product. Butyl octanoate was readily formed (98% in 20 h). Subsequently, ammoniolysis was performed. After 45 h of reaction the conversion of butyl octanoate was 95% (determined with HPLC), and octanamide and octanoic acid were formed in 85% resp 10% yield. After standard work-up procedures 76% octanamide (based on the amount of octanoic acid started with) was isolated. The formation of octanoic acid during the ammoniolysis is due to traces of water which are not removed by reflux over molecular sieves. We assumed that this water was strongly bound to the enzyme-preparation. To see whether this was caused by absorption of water by the enzyme or by the support, we carried out a large scale reaction with lipase SP525 (*Candida antarctica* B) immobilized on Accurel EP100 as catalyst, because this very apolar support (polypropene) would not absorb water.

In this case two equivalents of n-butanol were used in *t*-amyl alcohol as solvent, because the reaction rate of the ammoniolysis in pure n-butanol was to low to gain complete conversion in a reasonable reaction time. Again, the esterification was performed at diminished pressure under reflux over molecular sieves. After 24 hrs butyl octanoate was formed in 98% yield. Subsequent ammoniolysis yielded 97% of octanamide, and only traces of octanoic acid could be detected. After normal work-up procedures octanamide was isolated in 93% yield. Therefore, we can conclude that the formation of octanoic acid in the ammoniolysis is due to absorption of water on the support of lipase SP435 and that lipase SP525 immobilized on Accurel EP100 can best be used as catalyst in the one-pot synthesis.

When applied to chiral carboxylic acids, the one-pot synthesis is expected to take place with high enantioselectivity because enantioselective esterification and enantioselective ammoniolysis are combined in a double resolution process.

CONCLUSIONS

The lipases of *Candida antarctica* SP435 and SP525, *Humicola* SP398 and SP523 and *Pseudomonas alcaligenes* are able to perform ammoniolysis in a reaction medium of *t*-butyl alcohol saturated with ammonia. The lipases of *Rhizomucor miehei*, Amano PS, porcine pancreas and lipoprotein lipase from *Pseudomonas*, are able to perform ammoniolysis when an equimolar amount of NH_3 is used. The use of NH_4Cl and triethylamine as ammonia source is not suitable for most lipase preparations.

Ammoniolysis of non-chiral substrates (fatty esters) and chiral substrates (ibuprofen-2-chloroethyl ester, α-methylbenzyl alcohol) can be carried out. One of the great advantages of enzyme-catalyzed ammoniolysis is that the ammoniolysis is an order of magnitude more selective than corresponding enzymatic hydrolysis and alcoholysis reactions.

An efficient one-pot procedure from octanoic acid to octanamide via butyl octanoate is developed. It is best performed with lipase SP525 on Accurel EP100 as catalyst, with two equivalents of butyl alcohol in *t*-amyl alcohol as solvent.

EXPERIMENTAL

^{1}H- and ^{13}C-NMR spectra were recorded in CDCl$_3$ with TMS as internal standard using a Varian VXR-400S spectrometer. Analytical HPLC was performed using a Waters 590 pump on a reversed-phase column (Novapak C$_{18}$) at ambient temperature and with an effluent flow of 1.0 ml/min, with detection on an Erma ERC-7510 RI detector. Chiral HPLC was performed on a chiral straight-phase column (Baker, Chiralcel OD) at ambient temperature and with an effluent flow of 0.5 ml/min, with detection on a Shimadzu SPD-6A variable wavelength detector at 220 nm. Analytical GC was performed with a Varian Star 3400 on a CP-Sil5 25 m x 0.32 mm df = 0.12 µm column. Chiral GC-analysis was performed with a Hewlett Packard 5890A on a Astec Chiraldex G-TA 20 m x 0.25mm column. Melting points are uncorrected and were measured on a Büchi 510 melting-point apparatus.

All solvents and reagents were of reagent grade and were dried on molsieves before use.

Immobilisation:

Immobilisation of the enzymes on Accurel EP100 was performed according to a published procedure[8].

Table 4. *Enzyme preparations tested for ammoniolytic activity.*

Enzyme	carrier	pH-optimum	Supplier
lipase SP435 (*Candida antarctica*)	polyacrylate	5-9	Novo Nordisk A/S
lipase SP525 (*Candida antarctica*)	Accurel EP100	5-9	Novo Nordisk A/S
Amano PS lipase	Accurel EP100[a]	7	Amano Pharmaceuticals
Rhizomucor miehei lipase	anionic exchange resin	5.5-7.5	Novo Nordisk A/S
Rhizomucor miehei lipase	Accurel EP100[a]	5.5-7	Novo Nordisk A/S
Candida cylindracea lipase	not immobilized	7.7	Sigma
Candida cylindracea lipase	Accurel EP100[a]	7.7	Sigma
lipase SP398	Accurel EP100[a]	6-10	Novo Nordisk A/S
lipase SP523	Accurel EP100[a]	6-10	Novo Nordisk A/S
lipase porcine pancreas	not immobilized	7.7	Sigma
lipoprotein lipase from *Pseudomonas*	Accurel EP100[a]		Boehringer Mannheim GmbH
Rhizopus arrhizus lipase	Accurel EP100[a]	ca. 6	Boehringer Mannheim GmbH
lipase SP526	Accurel EP100[a]	5-9	Novo Nordisk A/S
lipase from *Pseudomonas alcaligenes*	Accurel EP100[a]	8-11	Gist-brocades
cholesterol esterase from *Candida cylindracea*	Accurel EP100[a]		Boehringer Mannheim GmbH
carboxylesterase NP	Accurel EP100[a]	6-11	Gist-brocades
protease SP522	Accurel EP100[a]		Novo Nordisk A/S
protease from *Aspergillus saitoi*	Accurel EP100[a]	2.8	Sigma
protease from *Rhizopus*	Accurel EP100[a]	3.0	Sigma
thermolysin	Accurel EP100[a]		DSM
subtilisin SP458	Accurel EP100[a]	3.8-8.0	Novo Nordisk A/S
subtilisin Carlsberg	Accurel EP100[a]	7.5	Sigma

[a] The enzymes were immobilized on Accurel EP100 according to a published procedure (Pedersen and Eigtved).

Screening experiments:

NH₃-saturated:

A solution of 5 ml of ammonia saturated t-butyl alcohol (12.5 mmol NH_3), ethyl octanoate, 500 µl, 2.5 mmol) and diethyleneglycol dibutyl ether (100 µl, internal standard) was shaken at 40°C. 100 mg of enzyme preparation was then added. The reaction was monitored with HPLC, Novapak C18 (eluent: 75/25 methanol/water, 0.01M NaOAc/HOAc pH 4.5). Results are given in Table 5.

One equivalent of NH₃:

A solution of 4 ml of t-butyl alcohol, 1 ml of ammonia-saturated t-butyl alcohol (2.5 mmol NH_3), ethyl octanoate (500 µl, 2.5 mmol) and diethyleneglycol dibutyl ether (100 µl, internal standard) was shaken at 40°C. 100 mg of enzyme preparation was then added. The reaction was monitored with HPLC, Novapak C18 (eluent: 75/25 methanol/water, 0.01M NaOAc/HOAc pH 4.5). Results are given in Table 5.

NH₄Cl/Et₃N:

A solution of 5 ml of t-butyl alcohol, ethyl octanoate (2.5 mmol), NH_4Cl (7.5 mmol), Et_3N (7.5 mmol) and diethyleneglycol dibutyl ether (100 µl, internal standard) were shaken at 40°C. 100 mg of enzyme preparation was added. The reaction was monitored with HPLC, Novapak C18 (eluent: 75/25 methanol/water, 0.01M NaOAc/HOAc pH 4.5). Results are given in Table 5.

Ammoniolysis of ethyl octanoate and ethyl butyrate:

A solution of 5 ml of t-butyl alcohol, 0.5 ml of ethyl octanoate or ethyl butyrate and 100 µl of diethyleneglycol dibutyl ether (internal standard) was stirred at room temperature. Ammonia was bubbled through the solution and 50 mg of enzyme preparation was added. The reaction was monitored with HPLC, Novapak C18 (eluent: 75/25 methanol/water, 0.01 M NaOAc/HOAc pH 4.5). Octanamide: yield (HPLC): 95% in 24 h, mp: 103.8-104°C, lit. 106-110°C[9].
Butanamide: yield (HPLC): 95% in 24 h, mp: 114-115°C, lit. 115-116°C[9].

Table 5. *Ammoniolytic activity of tested enzyme preparations at different ammonia concentrations.*

Enzyme preparation	Conversion[a] of ethyloctanoate (%) in 17 h, unless described otherwise		
	NH_3-saturated	Equimolar NH_3	NH_3 from NH_4Cl and triethylamine
C. antarctica lipase SP435	95 in 24 h	43	47
C. antarctica lipase SP525	100 in 24 h	55 in 24 h	87 in 24 h
Humicola lipase SP398	60 in 24 h	23	6
Humicola lipase SP523	85 in 24 h	38	13
Pseudomonas alcaligenes	95 in 24 h	53 in 24 h	26 in 24 h
Pseudomonas lipoprotein lipase	15 in 24 h	34	11
Rhizopus arrhizus lipase	10 in 72 h	8	1
C. cylindracea cholesterol esterase	5 in 72 h	0	0
Amano PS lipase	0	7	1
Rhizomucor miehei lipase	0	20	1
porcine pancreas lipase	0	12	4
SP526	0	2	0

[a] The conversion was determined by HPLC

Ammoniolysis of trilaurin

To a solution of trilaurin (5.00 g, 7.82 mmol) in ammonia-saturated *t*-butyl alcohol (50 ml, 125 mmol NH_3) Novozym 435 (250 mg) was added. The reaction mixture was shaken at 60°C during 24 h. The enzyme was filtered off and the solvent evaporated *in vacuo*. The resulting white solid was washed with water and dried *in vacuo*. Laurinamide: yield: 4.53 g, 22.8 mmol, 97%.

Ammoniolysis of esters of chiral carboxylic acids:

General experiment:

A solution of 5 ml of t-butyl alcohol, 200 µl of substrate and 100 µl of diethyleneglycol dibutyl ether (internal standard) was stirred at room temperature. Ammonia was bubbled through the solution and 50 mg of lipase SP435 was added. The conversion was monitored during the reaction with GC, column: CPSil5 25 m; The ee of the ester was monitored with chiral GC, column: Astec Chiraldex G-TA, 20 m.

Ammoniolysis of ethyl 2-chloropropanoate: conversion 49% in 30 min, ee_{ester} = 40%, E = 6.1

Ammoniolysis of ibuprofen 2-chloroethyl ester (3-(4-isopropylphenyl)-2-methylethanoic acid)

A solution of 5 ml of t-butyl alcohol, 0.5 ml of ibuprofen 2-chloroethyl ester, 100 µl of diethyleneglycol dibutyl ether was stirred at room temperature. Ammonia was bubbled through the solution and 50 mg of lipase SP435 was added. The reaction was monitored with HPLC: conversion (Novapak C18): eluent 60/40 acetonitrile/water, 0.01 M NaOAc/HOAc pH 4.3; ee (Chiralcel OD): eluent 85/15 hexane/2-propanol. ee_{ester} was measured, ee_{amide} ((R)-enantiomer) was calculated: ee_{amide} = ee_{ester}(1-conv)/conv.

After 48 h (conversion 56%, ee_{ester} 96% ((S)-enantiomer), E = 28) the enzyme was filtered off, and the solvent was evaporated *in vacuo*. Ibuprofen amide was crystallized from petroleum ether 40/60 at -20°C.

Ibuprofen amide: mp: 126.2 -126.9°C , [α] = -25.2 ° (ethanol) c=1.

^{1}H-NMR: δ 0.88, d, 6H, J = 6.6, $(CH_3)_2CH_2$; 1.42, d, 3H, J = 7.2, CH_3CH; 1.83, m, 1H, $CH(CH_3)_2$; 2.44, d, 2H, J = 7.2, $CH_2C_6H_6$; 3.63, q, 1H, J = 7.1, $CHCONH_2$; 7.09, d, 2H, J = 8.2, Ar-H; 7.24, d, 2H, J = 8.0, Ar-H.

^{13}C-NMR: δ 18.88, 22.72, 31.49, 46.07, 46.78, 128.19, 130.29, 140.41, 141.53, 180.23.

Hydrolysis of ibuprofen 2-chloroethyl ester and ethyl 2-chloropropanoate

To a mixture of 5 ml of dry *t*-butyl alcohol, 270 µl of water, 100 µl of diethyleneglycol dibutyl ether (internal standard) and 400 µl of ibuprofen 2-chloroethyl ester or 200 µl of ethyl 2-chloropropanoate 50 mg of lipase SP435 was addded. The reaction was followed with HPLC (ibuprofen 2-chloroethyl ester) conversion (Novapak C18): eluent 60/40 acetonitrile/water, 0.01 M NaOAc/HOAc pH 4.3; ee (Chiralcel OD): eluent 85/15 hexane/2-propanol; or GC (Astec Chiraldex G-TA) for ethyl 2-chloropropanoate. ee_{ester} was measured, ee_{acid} was calculated: $ee_{acid} = ee_{ester}(1-conv)/conv$.

Hydrolysis of ibuprofen 2-chloroethyl ester: conversion 63% in 24 h, ee_{ester} = 58%, (*S*)-enantiomer; E = 3.5.

Hydrolysis of ethyl 2-chloropropanoate: conversion 45% in 5 h, ee_{ester} = 31%, E = 2.7

Esterification of ibuprofen with 2-chloroethanol

To a reaction mixture containing 5 ml of *t*-butyl alcohol, 500 mg of ibuprofen, 3 ml of 2-chloroethanol, 100 µl of diethyleneglycol dibutyl ether (internal standard) and 900 mg zeolite (4A) to absorb the water formed, 50 mg of lipase SP435 was added. The reaction was shaken at 40°C, and monitored with HPLC (eluent as used for ammoniolysis).

conversion: 19% in 60 h, ee_{ester} = 75%, E = 8

Alcoholysis of ibuprofen 2-chloroethyl ester

To a solution of 5 ml of *t*-butyl alcohol, 1 ml of n-butyl alcohol, 500 µl of ibuprofen 2-chloro ethyl ester and 75 µl of dimethoxybenzene (internal standard) 50 mg of lipase SP435 was added. The reaction mixture was stirred with an overhead stirrer at room temperature. The reaction was monitored with HPLC (Chiralcel OD, eluent 83/17 hexane/ 2-propanol).

conversion: 56% in 45 h, $ee_{chloroethylester}$ = 86%, E = 16

Ammoniolysis of α-methylbenzyl n-butyrate :

A solution of 5 ml of t-butyl alcohol, 200 μl of α-methylbenzyl n-butyrate, 100 μl of diethyleneglycol dibutyl ether (internal standard) was stirred at room temperature. Ammonia was bubbled through the solution and 100 mg of lipase SP435 was added. The reaction was monitored with GC; ee of α-methylbenzyl alcohol was measured (Astec Chiraldex G-TA).

conversion 45% (120 h), ee of α-methylbenzyl alcohol = 98%,

(R)-enantiomer; E >100.

Transesterification of α-methylbenzyl n-butyrate:

A solution of 5 ml of t-butyl alcohol, 200 μl of α-methylbenzyl n-butyrate, 100 μl of diethyleneglycol dibutyl ether (internal standard) and 1 ml of absolute ethanol was stirred at room temperature. The reaction was monitored with GC; ee of α-methylbenzyl alcohol was measured (Astec Chiraldex G-TA).

conversion 46% (24 h), ee of α-methylbenzyl alcohol = 96%, (R)-enantiomer; E > 100.

Hydrolysis of α-methylbenzyl n-butyrate

To a solution of 5 ml of t-butyl alcohol, 200 μl of α-methylbenzyl n-butyrate, 100 μl of diethyleneglycol dibutyl ether (internal standard) and 270 μl of water 100 mg of lipase SP435 was added. The reaction mixture was shaken at room temperature. The course of the reaction was monitored with GC.

ee of α-methylbenzyl alcohol was measured (Astec Chiraldex G-TA).

conversion 47% (24 h), ee of α-methylbenzyl alcohol = 96%, (R)-enantiomer; E > 100.

Consecutive esterification and ammoniolysis

HPLC-analysis: column: Novapak C18, eluent: 80/20 methanol/water, pH 4.3.

Ammoniolysis of butyl octanoate in t-butyl alcohol/n-butyl alcohol mixtures.

To a solution of butyl octanoate (0.5 ml) and diethyleneglycol dibutyl ether (50 µl, internal standard) in *t*-butyl alcohol/n-butyl alcohol mixtures (5/0, 3/2, 1/4 and 0/5) was added lipase (50 mg). The reaction mixture was shaken at 40°C and the course of the reaction was followed with HPLC. The lipase-preparations used are: Lipase SP435 (*C. antarctica* immobilized), lipase SP525 (*C. antarctica*) immobilized on Accurel EP100, lipase SP523 (*Humicola*) immobilized on Accurel EP100, lipase from *Pseudomonas alcaligenes* immobilized on Accurel EP100 and lipase from *Rhizomucor miehei* on anionic resin. For results see Table 6.

Esterification with ethanol followed by ammoniolysis in t-butyl alcohol

To a solution of octanoic acid (0.5 ml, 3.1 mmol) and diethyleneglycol dibutyl ether (100 µl, internal standard) in ethanol (5 ml) lipase SP435 (100 mg) was added. The reaction mixture was shaken at 40°C for 24 h. Then, the enzyme was filtered off and the ethanol evaporated *in vacuo*. *t*-Butyl alcohol (5 ml) and fresh lipase SP435 (100 mg) were added and the reaction mixture was saturated with ammonia. The reaction mixture was shaken at 40°C for 24 h.

HPLC-yields: ethyl octanoate: 98% (24 h), octanamide 80% (24 h).

Esterification and ammoniolysis in n-butyl alcohol

To a solution of octanoic acid (5 ml) in n-butyl alcohol (50 ml) lipase SP435 (250 mg) was added. The reaction mixture was stirred for 24 h under reflux at 50°C *in vacuo*, and the refluxing n-butyl alcohol was dried over molecular sieves. Then the solution was saturated with ammonia. The course of the reaction was followed with HPLC.

HPLC-yields: butyl octanoate: 98% (24 h).

Under these reaction conditions ammoniolysis was not observed.

Table 6. *Ammoniolytic activity of lipase-preparations in mixtures of t-butyl alcohol and n-butyl alcohol.*

lipase-preparation	t-BuOH/ n-BuOH	conversion (%) in 24 h
SP435 on EP100	100/0	77
	50/50	9
	0/100	5
SP525 on EP100	100/0	99
	50/50	42
	0/100	13
SP523 on EP100	100/0	41
	50/50	27
	0/100	8
P. alcaligenes on EP100	100/0	80
	50/50	23
	0/100	11
Rh. miehei on anionic resin, 1 eq NH$_3$	100/0	15
	50/50	0
	0/100	0

Esterification and ammoniolysis in t-amyl alcohol

Lipase SP435: To a solution of octanoic acid (5 ml, 32 mmol) in *t*-amyl alcohol (50 ml) and n-butyl alcohol (3 ml, 32 mmol) was added lipase SP435 (250 mg). The reaction was stirred for 24 h under reflux at 50°C *in vacuo*. The refluxing *t*-amyl alcohol was dried over molecular sieves. Then, the reaction mixture was saturated with ammonia and stirred for 48 h at 40°C. The course of the reaction was followed with HPLC. After 48 h, the enzyme was filtered off and the solvent evaporated *in vacuo*. The product was crystallized from methanol/petroleum ether 40-60.

HPLC-yields: butyl octanoate: 98% in 24 h (70% in 2 h), octanamide: 85% in 48 h. Octanamide (isolated yield): 3.43 g (24 mmol, 76%).

102

Lipase SP525 on Accurel EP100: To a solution of octanoic acid (5 g, 35 mmol) in *t*-amyl alcohol (50 ml) and n-butyl alcohol (6 g, 81 mmol) was added lipase SP525 immobilized on Accurel EP100 (250 mg). The reaction was stirred for 24 h under reflux at 50°C *in vacuo*. The refluxing *t*-amyl alcohol was dried over molecular sieves. Then, the reaction mixture was saturated with ammonia and stirred for 168 h at 40°C. The course of the reaction was followed with HPLC. After 168 h, the enzyme was filtered off and the solvent evaporated *in vacuo*. HPLC-yields: butyl octanoate: 98% in 24 h, octanamide: 89% in 6 h, 97% in 168 h.

Octanamide (isolated yield): 4.62 g (32.3 mmol, 93%). m.p.: 104°C, lit. 106-110°C, 99% pure on HPLC.

REFERENCES

1. A. M. Klibanov, *Acc. Chem. Res.*, **23**, 114 (1990).

2. F. Björkling, H. Frykman, S. E. Godtfredsen and O. Kirk, *Tetrahedron*, **22**, 4587 (1992); M. C. de Zoete, F. van Rantwijk, L. Maat and R. A. Sheldon, *Recl. Trav. Chim. Pays-Bas*, **112**, 462 (1993).

3. V. Gotor, R. Brieva, C. Gonzalez and F. Rebolledo, *Tetrahedron*, **47**, 9207 (1991).

4. C. Astorga, F. Rebolledo and V. Gotor, *Synthesis*, 287 (1993).

5. A. L. Margolin, D.-F. Tai and A. M. Klibanov, *J. Am. Chem. Soc.*, **109**, 7885 (1987); A. L. Margolin and A. M. Klibanov, *J. Am. Chem. Soc.*, **109**, 3802 (1987).

6. V. Gotor in "Microbial Reagents in Organic Synthesis", S. Servi (ed), Kluwer Academic Publishers, Dordrecht, The Netherlands, pp. 199-208 (1992); H. Kitaguchi, F. A. Fitzpatrick, J. E. Huber and A. M. Klibanov, *Ann. N. Y. Acad. Sci.*, **613**, 656 (1990).

7. B. Tuccio, E. Ferré and L. Comeau, *Tetrahedron Lett.*, **32**, 2763 (1991).

8. S. Pedersen and P. Eigtved, WO 90/15868, 27 December 1990, Novo Nordisk A/S (1990).

9. R. C. Weast, Handbook of Chemistry and Physics, CRC Press, Inc, Florida, (1981).

Chapter 6

Enzymatic ammoniolysis of amino acid derivatives

Summary

The ammoniolysis of amino acid esters and derivatives was performed with lipases and proteases yielding the corresponding amino acid amides with moderate to high enantioselectivity. Phenylglycine methyl ester racemized slowly under the reaction conditions.

This chapter is in the press in *Recl. Trav. Chim. Pays-Bas* (1995).

INTRODUCTION

The use of enzymes in an unnatural environment is a valuable addition to the synthetic repertoire. Among the serine hydrolases, lipases are particularly versatile catalysts because of their ability to effect acyl-transfer reactions with a broad spectrum of nucleophiles[1]. Hydrogen peroxide[2], amines[3], hydrazines[4], and oximes[5] are among the compounds which act as acyl acceptors.

We[6] and others[7] recently reported that the lipase-catalyzed reaction of carboxylic esters and ammonia (ammoniolysis) afforded the corresponding amides in high yields under mild conditions. We also found that the reaction is enantioselective with respect to the acyl and alcohol moieties in the reactant; in some cases ammoniolysis proved to be up to an order of magnitude more enantioselective than the corresponding alcoholysis and hydrolysis reactions.

Developments in the manufacture of penicillin and cephalosporin antibiotics[8] prompted us to extend enzyme-catalyzed ammoniolysis to amino acid esters. In the future, D-phenylglycine amide and the corresponding 4-hydroxy compound could become important synthetic intermediates, because conventional chemical coupling procedures are being superseded by alternative, more environmentally acceptable enzymatic methods (see Scheme 1). Amino acid amides are also useful intermediates in peptide synthesis, because the C-terminus in many biologically active peptides consists of a primary amide group.

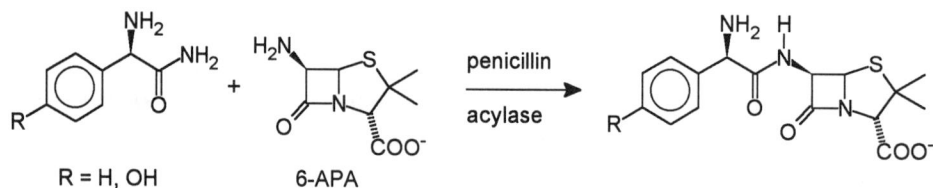

R = H, OH   6-APA

Scheme 1. *Enzyme-catalyzed coupling of D-(4-hydroxyphenyl)glycine amide with 6-aminopenicillic acid.*

DSM workers have described[9] the synthesis of D-phenylglycine amide by an L-specific aminopeptidase-catalyzed hydrolysis of the racemic amide which was prepared via a Strecker reaction from benzaldehyde (Scheme 2).

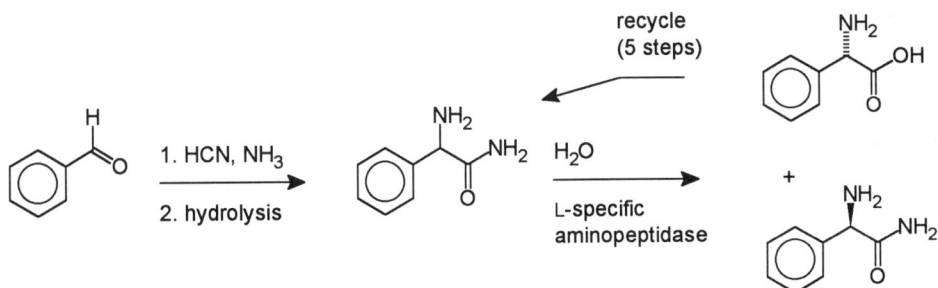

Scheme 2. *Resolution of D-phenylglycine amide.*

In the present paper we describe the lipase- and protease-catalyzed enantioselective ammoniolysis of amino acid esters as a potentially attractive alternative method for the preparation of enantiopure amino acid amides and their protected derivatives.

RESULTS AND DISCUSSION

Lipase-catalyzed ammoniolysis

A number of racemic amino acid esters were smoothly converted into the corresponding amides by *Candida antarctica* lipase SP435 (Table I). Protection of the amino group proved to be unnecessary; in fact Z-amino acid esters did not react, probably due to the bulk of the protecting group, with the exception of *N*-(benzyloxycarbonyl)alanine methyl ester (Z-Ala-OMe).

Phenylalanine methyl ester (Phe-OMe) and Z-Ala-OMe were converted to their corresponding amides by lipase SP435 with low enantioselectivity. *Humicola* lipase SP398 catalyzed the ammoniolysis of Phe-OMe at a lower rate, but with much better enantioselectivity; Z-Ala-OMe did not react. Because the enantiomers of the products could not be separated with chiral HPLC, the enantiomeric excesses of the starting materials were used to determine the enantiomeric ratio of these reactions. Phenylglycine methyl ester (Phg-OMe) was a poor substrate for all lipases tested[10] with the exception of *Candida antarctica* lipase SP435, which converted phenylglycine methyl ester to its D-(-)-amide with high enantioselectivity (enantiomeric excess, *ee*, 91% at 32% conversion). In this case, the enantiomeric excesses of both the starting material and the reaction product could be measured. It became clear that the starting material racemized slowly under the reaction conditions (5% in 24 h), but the product was stable. If, at a reaction time of 4 h (conversion 32%), the effect of racemisation of reactant is disregarded, the *ee* of the reactant (35%) corresponds with an *E* factor of 10, whereas the *ee* of the amide (91%) points to *E* 32. The amide was isolated from a preparative scale experiment as its Schiff base derivative in a yield of 32% with *ee* 92% (*E* 40). These results indicate an *E* factor of 35 ± 5.

Depending on the rate, *in situ* racemisation of the reactant could lead to increased yields of product up to a theoretical yield of 100%. Hence, we are currently investigating several methods in order to enhance the racemisation rate.

Table I. *Lipase-catalyzed ammoniolysis of amino acid esters[a].*

Substrate	Lipase	Reaction time (h)	Conv. (%)	ee_{ester} (%)	ee_{amide} (%)	E[11]
Phe-OMe	SP435	24	57	0	n.d.[b]	1
	SP398	23	25	25	n.d.[b]	9
Phg-OMe	SP435	4	32	35	91	35[c]
Z-Ala-OMe	SP435	22	18	8	n.d.[b]	2

[a] Conditions: see experimental section. [b] n.d.: not determined because enantiomers could not be separated with chiral HPLC. [c] See text.

Ammoniolysis with proteases

Since amino acids are natural substrates of proteases we reasoned that proteases should display catalytic activity in the ammoniolysis of amino acid esters. Indeed, four protease preparations were found to be active (Maxacal, Maxatase, SP539, subtilisin A) but a few conditions had to be fulfilled. Firstly, protection of the amino function was required, not only to prevent peptide formation, but also to increase activity, since proteases showed low activity towards unprotected amino acid esters. We assume that the extra amide bond which is created by protection of the amino function and which resembles a peptide bond, is required for proper binding of the substrate in the active site. Secondly, the concentration of ammonia was important. High ammonia concentrations inactivated the proteases, but when ammoniolysis was performed with two equivalents of ammonia (based on the amino acid ester) activity was retained. Finally, immobilisation of the proteases on Accurel EP100 resulted in an increased stability under ammoniolysis conditions.

While these investigations were underway a publication appeared on the ammoniolysis of amino acid derivatives catalyzed by alcalase[12]. NH_4Cl/Et_3N was used as the ammonia source and the protease applied was an unimmobilized preparation. Hence, we have compared the ammoniolysis of Z-Ala-OMe using this procedure with our results. It became evident that our procedure gave superior results with the tested proteases. We ascribe this to the stabilizing effect of immobilisation on Accurel EP100. The rapid deactivation of the proteases when using NH_4Cl/Et_3N as ammonia source is probably due to the basicity of Et_3N (Table II).

Table II. *Comparison of two methods of ammoniolysis of Z-Ala-OMe[a].*

Protease	Conversion (%) in 24 h, NH$_4$Cl/Et$_3$N, unimmobilized protease	Conversion (%) in 24 h, NH$_3$-saturated t-butyl alcohol, protease on Accurel EP100
SP 539	3	45
Subtilisin A	4	26
Maxacal	5	28
Maxatase	5	39

[a]Conditions: see experimental section.

Three racemic benzyloxycarbonyl protected amino acid esters (Z-amino acid esters) were subjected to protease-catalyzed ammoniolysis. Z-Phg-OMe was not accepted as substrate by any of the proteases tested, probably due to steric hindrance by the α-phenyl group. The L-enantiomers of the two other substrates were converted to their amides with high enantioselectivity (Table III), which enabled us to isolate the L-amides in high optical purity (see Experimental).

Other derivatives of amino acids

We also attempted the ammoniolysis of the hydantoin[a] and the oxazolinone derivatives of phenylalanine. These are interesting compounds because they racemize spontaneously under the reaction conditions and hence could afford 100% yield of a single enantiomer starting from a racemic mixture. However, neither lipases[13] nor proteases exhibited any reactivity towards the hydantoin. The oxazolinone was converted to the amide but the blank reaction in the absence of the enzyme was almost as fast as the enzymatic reaction.

---

[a] IUPAC name: imidazolidine-2,4-dione

Table III. *Protease-catalyzed ammoniolysis of amino acid esters[a].*

Substrate	Protease on Accurel EP100	Conversion (%) in 24 h	ee_{ester}[b] (%)	E[11]
Z-Phe-OMe	SP539	40	61	88
	Subtilisin A	37	52	22
	Maxacal	5	4	8
	Maxatase	27	35	54
Z-Ala-OMe	SP539	45	79	>>100
	Subtilisin A	26	35	>>100
	Maxacal	28	40	>>100
	Maxatase	39	64	>>100

[a] Conditions: see experimental section. [b] *ee*'s of the product could not be determined.

Summarizing, enzyme catalyzed ammoniolysis is an effective procedure to convert amino acid esters into the corresponding amides with high enantioselectivity. The possibility of *in situ* racemisation of amino acid esters such as phenylglycine methyl ester makes it, in principle, possible to obtain a complete conversion to one enantiomer of product starting from the racemic ester.

ACKNOWLEDGEMENTS

Generous gifts of enzymes by Novo-Nordisk A/S, Bagsvaerd, Denmark, are gratefully acknowledged. The authors also wish to thank Gist-brocades B.V. for a donation of Maxatase and Maxacal. We thank Dr. J. A. Peters and Mr. A. Sinnema for recording the 400-MHz NMR spectra.

EXPERIMENTAL

^{1}H- and ^{13}C-NMR spectra were recorded in DMSO or CDCl$_3$ with TMS as internal standard using a Varian VXR-400S, a Nicolet 200 MHz or a Varian T-60 spectrometer, as indicated. Analytical HPLC was performed using a Waters 6000A pump on a straight-phase column (Baker, Chiralcel OD column) at ambient temperature and with an effluent flow of 0.5 ml/min, with detection on a Shimadzu SPD-6A variable-wavelength detector at 254 nm, or on a reversed-phase column (Nucleosil C18) with an effluent flow of 1.0 ml/min with detection on an Erma ERC-7510 refractive-index detector. Optical rotations were measured on a Perkin-Elmer 241 polarimeter. Lipases SP435, SP398 and proteases SP539, SP458 and subtilisin A were donated by Novo Nordisk A/S and Maxatase and Maxacal were obtained from Gist-brocades B.V. All solvents and reagents were of reagent grade and were dried on molsieves before use.

Synthesis of amino acid esters

Phenylglycine methyl ester hydrochloride.
Phenylglycine (50 g, 0.33 mmol) was dissolved in dry methanol (200 ml) and HCl was bubbled through the solution for 6 h at room temperature. The solution was stirred for an additional 24 h. The solvent was evaporated *in vacuo*. The product was purified by dissolving it in methanol and precipitation with diethylether; yield 57 g, 0.28 mol, 85%.

Phenylalanine methyl ester hydrochloride.
Phenylalanine (2.56 g, 15.5 mmol) was dissolved in cold, dry methanol (30 ml). Thionyl chloride (1.8 g, 15 mmol) was slowly added and the reaction stirred for 24 h at room temperature. The solvent was evaporated *in vacuo* and the product was purified by dissolving it in methanol and precipitation with diethylether; yield 3.14 g, 14.6 mmol, 94%.

Alanine methyl ester hydrochloride. This compound was prepared using the procedure for phenylalanine methyl ester hydrochloride. The following amounts were used: alanine (8.9 g, 0.1 mol), methanol (100 ml) and thionyl chloride (7.25 ml, 0.1 mol); yield 13.5 g, 0.097 mol, 97%.

Z-amino esters were prepared by acylation of the amino acid esters with benzyloxycarbonyl chloride[14].

Authentic samples of amino acid amides were synthesized chemically by ammoniolysis of the corresponding amino acid esters[15].

Lipase-catalyzed ammoniolysis

Phenylalanine methyl ester hydrochloride. To this compound (200 mg, 0.93 mmol) was added 5 ml of ammonia-saturated *t*-butyl alcohol (12.5 mmol NH_3), 1,3-dimethoxybenzene (50 µl, internal standard) and lipase SP435 (50 mg) or lipase SP398 (50 mg). The reaction mixture was shaken at 40°C and monitored with HPLC (Chiralcel OD, eluent: hexane/propan-2-ol, 97/3, UV 254 nm). Lipase SP435 and SP398 afforded 57% conversion (ee_{ester} 0%) and 25% conversion (ee_{ester} 25%; E 9) in 24 h and 23 h, respectively.

Phenylglycine methyl ester hydrochloride. To this compound (200 mg, 0.99 mmol) was added ammonia-saturated *t*-butyl alcohol (5 ml, 12.5 mmol NH_3), 1,3-dimethoxybenzene (50 µl, internal standard) and lipase SP435 (50 mg). The reaction mixture was shaken at room temperature and monitored with HPLC (Chiralcel OD, eluent: hexane/propan-2-ol, 70/30, UV 254 nm). 32% conversion was observed in 4 h (ee_{amide} 91%; E 32).

Racemisation of phenylglycine methyl ester. To the hydrochloride (100 mg, 0.5 mmol) was added ammonia-saturated *t*-butyl alcohol (5 ml, 12.5 mmol NH_3) and 1,3-dimethoxybenzene (20 µl, internal standard). The reaction mixture was shaken at room temperature and monitored with HPLC (Chiralcel OD, eluent: hexane/propan-2-ol, 70/30, UV 254 nm). 5% racemisation to L-phenylglycine methyl ester was observed in 24 h.

N-(Benzyloxycarbonyl)alanine methyl ester hydrochloride. To this compound (100 mg, 0.4 mmol) was added ammonia-saturated *t*-butyl alcohol (5 ml, 12.5 mmol NH_3), 1,3-dimethoxybenzene (50 µl, internal standard) and lipase SP435 (50 mg). The reaction mixture was shaken at 40°C and monitored with HPLC (Chiralcel OD, eluent: hexane/propan-2-ol, 85/15, UV 254 nm). 18% conversion was observed in 22 h (ee_{ester} 8%, E 2).

Protease-catalyzed ammoniolysis

General procedure. To a solution of *N*-(benzyloxycarbonyl)amino acid ester and 1,3-dimethoxybenzene (50 µl, internal standard) in *t*-butyl alcohol (5 ml) was added ammonia saturated *t*-butyl alcohol and protease on Accurel EP100. The reaction mixture was shaken at 40°C and monitored with HPLC (details are given in Table IV).

Table IV. Details of the protease-catalyse ammoniolysis of Z-amino acid esters.

substrate (mg)	t-BuOH/ NH$_3$ (μl)	protease (mg)	HPLC column, eluent
Z-Phg-OMe (100)	132	subtilisin A (20)	Nucleosil C18, MeOH/H$_2$O, 60/40, 0.1% TFA
	132	SP458 (50)	
Z-Phe-OMe (100)	200	subtilisin A (20)	Chiralcel OD hexane/2-propanol, 93/7, UV 254nm
	200	SP539 (20)	
	200	Maxatase (20)	
	200	Maxacal (20)	
Z-Ala-OMe (100)	200	subtilisin A (20)	Chiralcel OD, hexane/2-propanol, 85/15, UV 254 nm.
	200	SP539 (20)	
	200	Maxatase (20)	
	200	Maxacal (20)	

N-(Benzyloxycarbonyl)phenylalanine methyl ester and N-(Benzyloxy-carbonyl)alanine methyl ester. Conversions and corresponding *ee*'s are given in Table III in Results and discussion. The *ee* values of the amides could not be determined because the enantiomers coeluted on chiral HPLC.

Isolation of the optically active amino acid amides

D-Phenylglycine amide. The product was isolated as a Schiff's base of which the *ee* could be determined on a Chiralcel OD column.

Phenylglycine methyl ester hydrochloride (1 g, 4.96 mmol) was dissolved in *t*-butyl alcohol saturated with ammonia (25 ml, 62.5 mmol of NH_3). Lipase SP435 (250 mg) was added and the reaction was shaken at room temperature. After 5 h of reaction, methanol (20 ml) was added to the reaction mixture, the enzyme was filtered off and the solution was evaporated *in vacuo*. Water (5 ml) was added to the residue and the pH of the solution was adjusted to 10 with 1M KOH. Benzaldehyde (175 µl) was then added and the mixture was stirred overnight at room temperature. The precipitate was filtered and dried *in vacuo* at room temperature. Analysis: Chiralcel OD, eluent: hexane/propan-2-ol, 70/30, UV 254 nm. Yield of *N*-benzylidene-D-phenylglycine amide: 276 mg (1.58 mmol, 32 %), *ee*: 92%, m.p. 123°C. ^{13}C-NMR[16] (200 MHz), ($CDCl_3$): δ 76.929, 127.145, 127.887, 128.425, 128.662, 131.477, 135.249, 139.169, 163.209, 174.065.

L-Z-Alanine amide. Z-Ala-OMe (1 g, 4 mmol) was dissolved in *t*-butyl alcohol (8 ml) and ammonia saturated *t*-butyl alcohol (2 ml, 5 mmol of NH_3) was added. Maxatase on Accurel EP100 (100 mg) was added and the reaction was shaken for 96 h at room temperature (conversion 44%). The enzyme was filtered off and the filtrate evaporated *in vacuo*. The product was purified by column chromatography (eluent: CH_2Cl_2/CH_3OH, 90/10). Yield of L-Z-alanine amide: 326 mg (1.47 mmol, 37%), m.p. 129.0-130.5°C; lit.[12] 129-130°C; $[\alpha]_D^{20}$ -4.9° (*c* 2, methanol), lit.[12]$[\alpha]_D$ -4.5° (*c* 2, methanol). ^{1}H-NMR (400 MHz), ($CDCl_3$): δ 1.40, d, 3H, CH_3; 4.26-4.29, m, 1H, CHCH$_3$; 5.1, s, 2H, CH_2; 5.38, d, 1H, NH; 5.60 and 6.11, 2H, $CONH_2$; 7.35, m, 5H, C_6H_5. ^{13}C-NMR (400 MHz), ($CDCl_3$): δ 18.503, 50.130, 67.085, 128.033, 128.253, 128.564, 135.120, 156.072, 175.106.

L-Z-Phenylalanine amide

Z-Phe-OMe (1.25 g, 4 mmol) was dissolved in *t*-butyl alcohol (8 ml) and ammonia saturated *t*-butyl alcohol (2 ml, 5 mmol of NH_3) was added. SP539 on Accurel EP100 (100 mg) was then added and the reaction was shaken for 120 h at 40°C. The enzyme was filtered off and washed with methanol and the filtrate was evaporated *in vacuo*. The product was purified by column chromatography (eluent: CH_2Cl_2/CH_3OH, 99/1); yield 496 mg (1.66 mmol, 42%), m.p. 159--160.5°C, lit[12]. 164-165°C; $[\alpha]_D^{20}$ -5.8° (*c* 2, methanol), lit.[12] $[\alpha]_D^{25}$ -5.15° (*c* 2, methanol). ^{1}H-NMR (60 MHz), (CDCl3): δ 3.2, d, 2H, $C_6H_5C\underline{H}_2CH$; 4.4, m, 1H, $C\underline{H}CONH_2$; 5.1, s, 2H, $OCH_2C_6H_5$; 7.1-7.3, m, 10H, 2 C_6H_5. ^{13}C-NMR (200 MHz), (DMSO): δ 37.372, 55.948, 64.999, 126.042, 127.242, 127.480, 127.858, 128.098, 129.005, 136.877, 138.114, 155.659, 173.283.

REFERENCES

1. For a review see: M.C. de Zoete, F. van Rantwijk and R.A. Sheldon, *Catalysis Today,* **22**, 563 (1994)..

2. F. Björkling, S. E. Godtfredsen and O. Kirk, J. Chem. Soc., Chem. Commun., 1301 (1990), F. Björkling, H. Frykman, S. E. Godtfredsen, and O. Kirk, *Tetrahedron* **48**, 4587 (1992).

3. H. Kitaguchi, P. A. Fitzpatrick, J. E. Huber and A. M. Klibanov, *J. Am. Chem. Soc.* **111**, 3094 (1989); Z. Djeghaba, H. Deleuze, B. De Jeso, D. Messadi and B. Maillard, *Tetrahedron Letters* **32**, 761 (1991); M. Quiros, V. M. Sanchez, R. Brieva, F. Rebolledo and V. Gotor, *Tetrahedron: Asymmetry* **4**, 1105 (1993); S. Puertas, R. Brieva, F. Rebolledo and V. Gotor, *Tetrahedron* **49**, 4007 (1993); V. Gotor, R. Brieva, C. Gonzalez and F. Rebolledo, *Tetrahedron* **47**, 9207 (1991).

4. C. Astorga, F. Rebolledo and V. Gotor, *Synthesis* 287 (1993); C. Astorga, F. Rebolledo and V. Gotor, *Synthesis* 350 (1991).

5. V. Gotor, C. Astorga, F. Rebolledo and E. Menedez, *Indian Journal of Chemistry* **31B**, 906 (1992); V. Gotor, "Microbial Reagents in Organic Synthesis", S. Servi, ed., Kluwer Academic Publishers, Dordrecht, The Netherlands, (1992) p199.

6.   M.C. de Zoete, A.C. Kock-van Dalen, F. van Rantwijk and R.A. Sheldon, *J. Chem. Soc., Chem. Commun.* 1831 (1993); M.C. de Zoete, A.C. Kock-van Dalen, F. van Rantwijk and R.A. Sheldon, *Biocatalysis* **10**, 307 (1994).

7.   M. J. Garcia, F. Rebolledo and V. Gotor, *Tetrahedron Letters* **34**, 6146 (1993).

8.   S. G. Kaasgaard and U. Veitland, Novo Nordisk A/S, PCT Int. Appl. WO 92 01,061, 23 Jan. 1992. [C. A. 116: 150153e], N. K. Maladkar, *Enzyme Microb. Technol.* **16**, 715 (1994).

9.   R. A. Sheldon, H. E. Schoemaker, J. Kamphuis, W. H. J. Boesten and E. M. Meijer in: "Stereoselectivity of pesticides: Biological and chemical problems", E. Ariëns, J. J. S. van Rensen and W. Welling, eds., Elsevier, Amsterdam, 1988, p. 409 and references cited therein.

10.   The lipase preparations tested were: Lipase from Amano PS, lipase from *Rhizopus arrhizus* on Accurel EP100, lipase from *Mucor miehei* on anionic resin, porcine pancreas lipase unimmobilized, lipoprotein lipase from *Pseudomonas* on Accurel EP100, *Humicola* lipase SP398 on Accurel EP100, *Candida antarctica* lipase SP435 immobilized.

11.   For explanation and calculation of E: K. Faber, "Bio-Transformations in Organic Chemistry", Springer-Verlag Berlin Heidelberg, Germany (1992).

12.   S-T. Chen, M-K. Jang and K-T. Wang, *Synthesis* 858 (1993).

13.   The inability of lipases to convert hydantoins was also reported by J. Z. Crich, R. Brieva, P. Marquart, R-L Gu, S. Flemming and C. J. Sih, *J. Org. Chem.* **58**, 3252 (1993).

14.   J. B. West and C.H. Wong, *J. Org. Chem.* **51**, 2728 (1986).

15.   H. Reilen and Knöpfe, *J. Lieb. Ann. Chem.*, **523**, 199 (1936).

16.   Our data are in agreement with previous results: private communication with dr. B. Kaptein, DSM Research.

Summary

Lipase- and protease-catalyzed transformations with unnatural acyl acceptors

This thesis is devoted to enzyme-catalyzed transformations with unnatural acyl acceptors. It describes the use of hydrolases (especially lipases and proteases) in organic solvents with hydrogen peroxide and ammonia as acyl acceptors.

With hydrogen peroxide as acyl acceptor (Chapter 2, 3 and 4) peroxycarboxylic acids are formed under mild reaction conditions, which can be used subsequently in oxidation reactions.

The hydrolase-catalyzed ammoniolysis (in which ammonia acts as acyl acceptor) is a new enzymatic reaction (Chapter 5 and 6). It can be applied in the (enantioselective) synthesis of amides under mild reaction conditions.

Chapter 1 provides a review on lipase-catalyzed transformations. Hydrolysis of unnatural substrates, esterification and transesterification reactions are discussed. Much attention is paid to the use of unnatural acyl acceptors such as hydrogen peroxide, sugars, amines, ammonia and hydrazines. The use of unnatural acyl donors is only shortly described.

In chapter 2 the lipase-catalyzed formation of peroxyoctanoic acid is described. This model reaction was studied in various solvents and with several lipase preparations as catalyst. The effect of the reaction temperature was also established. It appeared that the initial rate of formation increased with increasing reaction temperature and was highest with lipase SP435 (from *Candida antarctica*) in acetonitrile as solvent.

Chapter 3 deals with the lipase-catalyzed formation of chiral peroxycarboxylic acids and subsequent epoxidations. It was estimated that chiral peroxycarboxylic acids could induce enantioselectivity in epoxidation reactions. Although chiral peroxycarboxylic acids were formed in acetonitrile and subsequent epoxidation took place, induction of enantioselectivity was not observed.

Chapter 4 describes an application of the lipase-catalyzed *in situ* formation of peroxycarboxylic acids. Penicillin G was oxidized to penicillin G sulfoxide in high yield with enzymatically generated peroxyoctanoic acid. This enzymatic procedure allows the use of a catalytic amount of carboxylic acid, thus circumventing the use

119

of isolated peroxyacetic acid, which is explosive even at low temperatures. Moreover, a procedure in which penicillin G was oxidized with hydrogen peroxide at low reaction temperature is described.

The subject of chapter 5 is a new enzymatic reaction: the ammoniolysis of carboxylic esters. It can be used for the synthesis of amides under mild reaction conditions (atmospheric pressure and a reaction temperature of about 40°C). Fatty amides were easily prepared. Furthermore, ammoniolysis proved to be an efficient way to perform optical resolution of esters derived from chiral carboxylic acids, because it showed higher enantioselectivity than conventional enzymatic methods such as hydrolysis and esterification. Ammoniolysis could also be used for the resolution of esters derived from chiral alcohols. The development of a "one-pot procedure" which combines lipase-catalyzed (enantioselective) esterification with ammoniolysis, resulted in a new method to obtain (chiral) carboxylic amides starting from the parent carboxylic acids with high (enantio-) selectivity.

In chapter 6 the use of lipases and proteases in the synthesis of amino acid amides from amino acid esters and derivatives is reported. Lipases are able to catalyze ammoniolysis of unprotected amino acid esters. The ammoniolyis of phenylglycine methyl ester.HCl yielding (*D*)-phenylglycine amide is catalyzed by *Candida antarctica* lipase SP435 with high enantioselectivity. Proteases, on the contrary, require a protecting group on the amino function to perform ammoniolysis. Unprotected amino acid esters are not accepted as substrates. The proteases show, however, excellent enantioselectivity.

Marian de Zoete

Samenvatting

Lipase- en protease-gekatalyseerde omzettingen met niet-natuurlijke acylacceptoren

Dit proefschrift is gewijd aan enzym-gekatalyseerde omzettingen met niet-natuurlijke acylacceptoren. Het beschrijft het gebruik van hydrolasen, in het bijzonder van lipasen en proteasen, in organisch milieu met waterstofperoxide en ammoniak als acylacceptoren.

Wanneer waterstofperoxide gebruikt wordt als acylacceptor (Hoofdstuk 2, 3 en 4) worden er peroxycarbonzuren gevormd onder milde reactie-omstandigheden. Deze verbindingen kunnen vervolgens gebruikt worden als oxidant in oxidatiereacties.

De in hoofdstuk 5 en 6 beschreven ammoniolyse met hydrolasen is een nieuwe enzymatische reactie. De reactie kan worden toegepast voor de (enantioselectieve) synthese van amides onder milde reactie-omstandigheden.

In hoofdstuk 1 wordt een overzicht gegeven van lipase-gekatalyseerde omzettingen, waaronder bijvoorbeeld hydrolyse van niet-natuurlijke substraten, veresterings- en transveresteringsreacties. Er wordt veel aandacht besteed aan het gebruik van niet-natuurlijke acylacceptoren, zoals waterstofperoxide, suikers, amines, ammoniak en hydrazines. Het gebruik van niet-natuurlijke acyldonoren wordt slechts aangestipt.

Hoofdstuk 2 beschrijft de lipase-gekatalyseerde vorming van peroxyoctaanzuur. Deze modelreactie is bestudeerd in verschillende oplosmiddelen en met verscheidene lipase-preparaten als katalysator. Ook werd het effect van de reactietemperatuur bepaald. Er is gebleken dat de initiële reactiesnelheid toeneemt bij hogere temperatuur. De hoogste initiële reactiesnelheid werd gemeten in acetonitril met lipase SP435 (van *Candida antarctica*) als katalysator.

Hoofdstuk 3 gaat over de vorming van chirale peroxycarbonzuren en hun gebruik in epoxidatiereacties. Er werd aangenomen dat chirale peroxycarbonzuren enantioselectiviteit zouden kunnen induceren in epoxidatiereacties. Alhoewel chirale peroxycarbonzuren gevormd werden in acetonitril en er vervolgens ook epoxidatie optrad, werd er geen inductie van enantioselectiviteit waargenomen.

Hoofdstuk 4 behandelt een toepassing van de lipase-gekatalyseerde *in situ* vorming van peroxycarbonzuren. Hierbij werd penicilline G in hoge opbrengst geoxideerd tot

121

penicilline G sulfoxide door enzymatisch gevormd peroxyoctaanzuur. Bij deze enzymatische methode wordt uitgegaan van een katalytische hoeveelheid carbonzuur, waarbij dus het gebruik van stoichiometrische hoeveelheden peroxyazijnzuur, dat zelfs bij lage temperatuur al explosief is, wordt vermeden.

Het onderwerp van hoofdstuk 5 is een nieuwe enzymatische reactie, namelijk de ammoniolyse van carbonzure esters. Deze reactie kan worden toegepast in de synthese van amides onder milde reactie-omstandigheden (atmosferische druk en een reactietemperatuur rond 40°C). Op deze manier werden vetzuuramiden in hoge opbrengst gesynthetiseerd. Bovendien kan de reactie worden gebruikt voor optische resolutie. Er is gebleken dat ammoniolyse een betere methode is voor de optische resolutie van esters van chirale carbonzuren dan conventionele methoden als enzymatische hydrolyse of verestering, aangezien ammoniolyse selectiever is. Ammoniolyse bleek ook een goede methode voor de optische resolutie van esters van chirale alcoholen.

De ontwikkeling van een één-pots reactie, waarin lipase-gekatalyseerde verestering gecombineerd wordt met ammoniolyse, heeft een nieuwe methode opgeleverd voor de synthese van (chirale) amiden uitgaande van de corresponderende carbonzuren met hoge (enantio)selectiviteit.

In hoofdstuk 6 wordt in gegaan op het gebruik van lipasen en proteasen in de ammoniolyse van derivaten van aminozuren. Lipasen zijn in staat de ammoniolyse van onbeschermde aminozure esters te bewerkstelligen. Zo werd de ammoniolyse van de methyl ester van (D)-fenylglycine met hoge enantioselectiviteit gekatalyseerd door lipase SP435 (van *Candida antarctica*). Aan de andere kant vereisen de proteasen een beschermende groep op de aminofunktie, omdat onbeschermde aminozuren niet als substraat geaccepteerd worden. De proteasen vertonen echter in het algemeen een zeer goede enantioselectiviteit.

Marian de Zoete

Dankwoord

Op één van de laatste pagina's van mijn proefschrift wil ik alle mensen bedanken die een grote of kleine bijdrage hebben geleverd om mijn proefschrift zo te laten worden zoals het nu voor u ligt.

Allereerst wil ik natuurlijk mijn promotor Roger Sheldon bedanken voor zijn enthousiaste begeleiding en de onophoudelijke stroom aan ideeën waarvan er een paar hier in dit proefschrift zijn uitgewerkt. De vrijheid die ik heb gehad om de ideeën die mij het meest aanspraken er tussenuit te vissen en uit te werken heb ik altijd zeer gewaardeerd.

Natuurlijk wil ik ook mijn toegevoegd promotor Fred van Rantwijk bedanken voor de vele zinvolle discussies en suggesties op velerlei gebied. Wanneer ik niet zo'n "wetenschappelijke praatpaal" gehad zou hebben was mijn promotie-onderzoek vast een stuk minder vlot en soepel verlopen.

Leen Maat wil ik bedanken voor zijn begeleiding, vooral op het gebied van "het schrijven": zijn grondige correctie van maandverslagen en manuscripten heeft mij voor veel schrijffouten behoed.

Many thanks go to the workers of Novo-Nordisk: Leo Nieuwenhuis, Elie Berghmans, Lars Dalgårt Andersen, Allan Svendson, Steen Skjold-Jørgenson, Anne Mørkeberg-Larssen, Bent Riber Petersen and Kim Clausen. I thank them for providing us with all kinds of enzyme preparations and for the very fruitful discussions, which have contributed to a better understanding of the behaviour of enzymes in organic media.

De prettige samenwerking met Lida Kock-van Dalen op het onderwerp ammoniolyse heeft tot veel resultaten geleid. Zonder al haar inspanningen zou er veel niet zo snel afgerond zijn en hadden we vast het octrooi niet op tijd rond gekregen.

Pierre Peeters en Anne Ouwehand, die in het kader van hun studie bij mij stage hebben gelopen, hebben beide een grote bijdrage geleverd aan mijn proefschrift. Het pionierende werk van Pierre op het gebied van chirale peroxycarbonzuren staat gedeeltelijk in hoofdstuk 2 en hoofdstuk 3 beschreven. Het werk van Anne op het gebied van lipase-gekatalyseerde ammoniolyse van aminozuurderivaten heeft geleid tot een publikatie en staat beschreven in hoofdstuk 6.

Verder wil ik Adrie Knol-Kalkman, Loek van Leeuwen, Joop Peters, Anton Sinnema en Anton van Estrik bedanken voor al hun inspanningen op het gebied van de instrumentele analyse-technieken. Ernst Wurtz wil ik graag bedanken voor al zijn ondersteuning vooral bij de, soms ook erg smerige, proeven met ammoniak.

Mieke van der Kooij, de vraagbaak voor alle administratieve en andere rompslomp, wil ik bedanken voor al haar antwoorden op mijn nooit ophoudende stroom vragen op dat gebied.

Mijn collega's van de koffieclub van de T5-zaal hebben mijn promotietijd tot een onvergetelijke tijd gemaakt. De vele zinnige en zinloze discussies op allerlei gebied zal ik niet gauw vergeten. Een speciale plaats wil ik inruimen voor mijn mede- en exkamergenoten: Gert Barf, Marion van Deurzen, Erwin Mombarg en Bob de Goede. Dankzij jullie gezelligheid was de sfeer in onze kamer buitengewoon goed.

Een speciaal woord van dank gaat naar mijn ouders. Jullie warme belangstelling en waardering zijn voor mij een grote steun geweest.

Op de laatste maar eigenlijk de belangrijkste plaats wil ik mijn vriend en maker van de foto op de voorkant van dit boekje Mark Steverink bedanken. Lieve Mark, als leek in de scheikunde moet het vast niet al te gemakkelijk geweest zijn om al mijn verhalen te begrijpen. Toch hebben al jouw onzinnige maar vaak rake opmerkingen over mijn promotie-onderzoek mij altijd de zonnige kant van alles laten zien. Jouw steun en enthousiasme hebben mij veel vertrouwen in de toekomst gegeven.

124

Curriculum Vitae

Marian de Zoete werd geboren op 14 juni 1968 te 's-Gravenhage. In 1986 behaalde zij het gymnasium-β diploma aan het St.-Maartenscollege te Voorburg. In datzelfde jaar begon zij met haar studie Scheikunde aan de Rijksuniversiteit Leiden. Deze studie werd afgerond met een onderzoek naar de synthese en zure solvolyse van enolethers van glycol onder supervisie van prof. dr A. van der Gen. Het doctoraalexamen werd behaald in februari 1991.

Vanaf maart 1991 verrichtte zij een promotie-onderzoek bij de vakgroep Organische Chemie en Katalyse aan de Technische Universiteit Delft onder supervisie van prof. dr R. A. Sheldon en dr ir F. van Rantwijk. De resultaten van dit onderzoek zijn beschreven in dit proefschrift. Voor een deel van het promotie-onderzoek (de ammoniolyse) behaalde zij de DSM-prijs Chemie en Technologie 1994 (tweede prijs).

.